# SpringerBriefs in Stem Cells

For further volumes:
http://www.springer.com/series/10206

Lee Yee-Ki · Siu Chung-Wah

# Calcium Handling in hiPSC-Derived Cardiomyocytes

 Springer

Lee Yee-Ki
University of Hong Kong
Pokfulam
Hong Kong SAR

Siu Chung-Wah
University of Hong Kong
Pokfulam
Hong Kong SAR

ISSN 2192-8118    ISSN 2192-8126 (electronic)
ISBN 978-1-4614-4092-5  ISBN 978-1-4614-4093-2 (eBook)
DOI 10.1007/978-1-4614-4093-2
Springer New York Heidelberg Dordrecht London

Library of Congress Control Number: 2012940714

Printed on acid-free paper

Springer is part of Springer Science+Business Media (www.springer.com)

# Preface

Stem cells are proven to be able to differentiate into all types of cells in our bodies, including cardiomyocytes. In spite of the immuno-privileges of autologous transplantation of human induced pluripotent stem cells (hiPSCs), current safety issues include the potential risk of teratoma formation and immature phenotype of cardiomyocytes, leading to arrhythmia. This book gives an overview of the author's experience in the assessment of calcium homeostasis in hiPSC-derived cardiomyocytes (CMCs).

The first part of the book explains the sarcoplasmic reticulum (SR) as the potential target to assess maturity because of its crucial role in calcium ion flux during contraction. The second part suggests other protocols to assess the efficiency of calcium homeostasis in terms of SR junctional protein abundance and function, and the t-tubule structure of the ventricular-like cell should not be ignored. The third part of the book summarizes the comparison of calcium homeostasis of human embryonic stem cell – CMCs and hiPSC-CMCs. Last, the authors suggest a potential disease model to study the pathology of calcium handling-related defects.

The novelty of this book provides new insights in assessing stem cell – derived CMC maturity before clinical application. In addition to the therapeutic purposes of the cells, iPSC-CMCs also provide a new platform for pathological study as well as drug screening of cardiovascular diseases with calcium handling abnormalities.

# Acknowledgments

This Springer Brief, written by Dr. Lee Yee Ki and Dr. Siu Chung Wah, is based on the article "Calcium Homeostasis in Human Induced Pluripotent Stem Cell–Derived Cardiomyocytes," which appeared in the journal *Stem Cells Reviews and Reports* in 2011. Both principal authors were involved in designing the experiments and performing data analysis, and Dr. Lee, specifically, participated in all the experiments included in this project. Dr. Ng Kwong Man was involved in calcium confocal imaging, cardiac cell differentiation, and data analysis. Mr. Lai Wing Hon maintained the pluripotent stem cell culture and differentiated cardiomyocytes. Dr. Chan Yau Chi and Dr. Lau Yee Man took part in electrophysiology studies and cardiac cell isolation. Professor Tse Hung Fat also participated in designing the experiments and manuscript editing. Dr. Siu Chung Wah took part in experimental design and manuscript writing.

The article referenced above was the collaborative work of the following seven authors:

Lee Yee Ki, Ng Kwong Man, Lai Wing Hon, Chan Yau Chi, Lau Yee Man, Tse Hung Fat, and Siu Chung Wah.

All authors are affiliated with Cardiology Division, Department of Medicine, Queen Mary Hospital, The University of Hong Kong, China; and Research Center of Heart, Brain, Hormone and Healthy Aging, Li Ka Shing Faculty of Medicine, The University of Hong Kong, China.

Acknowledgments

# Contents

# Abbreviations

| | |
|---|---|
| ACE | Angiotensin-converting enzyme |
| AMI | Acute myocardial failure |
| BMP-4 | Bone morphogenetic protein-4 |
| CASQ2 | Calsequestrin |
| CHF | Chronic heart failure |
| CICR | Calcium-induced calcium release |
| CPVT | Catecholaminergic polymorphic ventricular tachycardia |
| CMC | Cardiomyocyte |
| DAD | Delayed afterdepolarization |
| Dkk-1 | Dickkopf-related protein 1 |
| EB | Embryonic body |
| EC | Excitation–contraction |
| END-2 | Endoderm-like cell |
| ESC | Embryonic stem cell |
| ESC-CMC | Embryonic stem cell–derived cardiomyocyte |
| HIF-1$\alpha$ | Hypoxia-inducible factor 1-alpha |
| hESC | Human embryonic stem cell |
| hiPSC | Human induced pluripotent stem cell |
| IP3R | Inositol-1,4,5-trisphosphate receptors |
| JUN | Junctin |
| KB | Kraftbrühe |
| MI | Myocardial infarction |
| NCX | $Na^+$–$Ca^{2+}$ exchanger |
| Oct4 | Octamer-binding transcription factor 4 |
| PCI | Percutaneous coronary intervention |
| PLB | Phospholamban |
| qPCR | Quantitative reverse transcription-polymerase chain reaction |
| RT-PCR | Reverse transcription-polymerase chain reaction |
| RyR2 | Ryanodine receptor-2 |

| SERCA2a | Sarcoplasmic reticulum calcium ATPase-2a |
| SOICR | Store overload-induced $Ca^{2+}$ release |
| SR | Sarcoplasmic reticulum |
| tPa | Tissue plasminogen activators |
| TR | Thyroid receptor |
| TRDN | Triadin |
| TRE | Thyroid response element |
| VEGF | Vascular endothelial growth factor |

# Chapter 1
# Calcium Handling in hiPSC-Derived Cardiomyocytes

**Abstract** Calcium is crucial in governing the contractile activities of myofilaments in cardiomyocytes. The characterization of calcium handling properties in human induced pluripotent stem cell–derived cardiomyocytes (iPSC-CMCs) is of significant interest and pertinent to the stem cell and cardiac regenerative field because of their potential patient-specific therapeutic use. In this book, readers can learn the approaches and parameters that have to be considered for calcium handling studies: these include the importance of the sarcoplasmic reticulum (SR), which governs the maturity of cardiomyocytes; the role of SR junctional proteins that facilitate calcium release and uptake through the SR; and the extent of immaturity of iPSC-CMCs compared with those derived from human embryonic stem cells in terms of calcium homeostasis. Future study would suggest potential uses of hiPSC-CMCs as a platform for cardiomyopathy disease modeling, especially certain cardiac defects related to impaired calcium homeostasis causing arrhythmia and pumping defeats. Last, a detailed protocol for calcium-sensitive dye calibration, calcium transient recording, and electrical contraction coupling is included.

## Myocardial Infarction and Heart Failure

Myocardial infarction (MI) is one of the most common causes of chronic heart failure (CHF) (Lloyd-Jones et al. 2010). The causative effect of MI on CHF is a complicated process with multiple factors known as left ventricular (LV) remodeling. MI is caused by interruption of the blood supply to certain regions of the heart, causing irreversible cardiomyocyte (CMC) death. This event is most commonly caused by occlusion of the epicardial coronary artery, followed by the rupture of a vulnerable atherosclerotic plaque in the artery, thus resulting in cardiac tissue ischemia with shortage of oxygen and nutrient supply. In the initial stage, the heart becomes hypertrophic to compensate for deteriorated cardiac function to maintain cardiac output and blood supply to vital organs. However, if the situation becomes

L. Yee-Ki and S. Chung-Wah, *Calcium Handling in hiPSC-Derived Cardiomyocytes*,    1
SpringerBriefs in Stem Cells, DOI 10.1007/978-1-4614-4093-2_1,
© Lee Yee-Ki and Siu Chung-Wah 2012

**Pluripotent Stem Cell for Cardiac Regeneration**

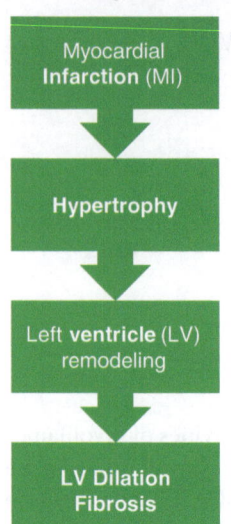

Current clinical therapeutic strategies:

• ACE inhibitor
• Thrombolytic agent e.g. tPA
• Beta-blocker
• Percutaneous coronary intervention (PCI)

Limited efficacy in aged patient

**Aim to investigate the potential to use pluripotent stem cell for cardiac regeneration:**

1) Regenerate electrical conduction :
   improvement of heart block

2) Reconstruct physical structure:
   proper mechanical contractile function

**Fig. 1.1** Complications of myocardial infarction and alternatives of current therapeutic strategies

worse, the consequence is postinfarction LV remodeling, which includes left ventricle dilation, fibrosis (scar-forming) impeding cardiac contractile function because of excessive cellular matrix buildup, and myocardial hypertrophy (Yousef et al. 2000).

CMCs are the cellular workhorse for cardiac contractile function. Their loss in myocardial infarction is irreversible because of their limited regenerative capacity, with a lack of stem cell pool, and often results in heart failure. The cardiac repair mechanism is a dynamic between cell apoptosis and progenitor cell proliferation. However, the ongoing cell loss becomes more severe in the aged myocardium. Despite recent advances in reperfusion therapy for acute myocardial infarction (AMI) and pharmacotherapy for post-MI LV remodeling, the incidence and mortality of post-MI heart failure are increasing (Velagaleti et al. 2008). As a result, there is a major unmet need for treatment of post-MI heart failure caused by progressive LV remodeling after the initial insult to the myocardium. Prompt reperfusion of the infracted artery is critical to prevent infarct expansion. The current pharmacological therapy for congestive heart failure mainly relies on angiotensin-converting enzyme (ACE) inhibitors, thrombolytic agents (tissue plasminogen activators, tPA), and beta-blockers (GUSTO Investigators 1993). The surgical interventions, percutaneous coronary intervention (PCI), attenuate myocardial necrosis and thus reduce mortality (Keeley et al. 2003) (Fig. 1.1). However, the foregoing various interventions and surgical therapies have limited efficacy in repairing or regenerating damaged myocardium.

Given the shortage of donor hearts for transplantation, recently there has been tremendous interest in developing novel stem cell therapies for the prevention and treatment of post-MI LV remodeling and dysfunction. However, in a typical MI involving 25% of the LV, there would be a loss of approximately $1 \times 10^9$ cardiomyocytes (CM), for which this intrinsic repair mechanism will not be able to restore the damaged myocardial contractility and to improve electrophysiological function by restoration of cardiac cell conductivity (Gepstein 2002; Nadal-Ginard et al. 2003), or even to mobilize and introduce exogenous stem cell to repair this deficit of CMs.

## Embryonic Stem Cells/Pluripotent Stem Cells

Embryonic stem cells (ESCs) originally have been derived from undifferentiated cells of the inner cell mass (ICM) of embryos at the blastocyst stage. The cells of ICM are pluripotent, meaning that they are able to develop into three germ layers. However, these differentiation processes occur only when ESCs are cultivated in the absence of leukemia inhibitory factor (LIF), and in the suspension culture in which they are grown to multicellular spheroidal tissues, termed *embryoid bodies* (EBs), they have been shown to differentiate spontaneously in vitro into cellular derivatives of all three primary germ layers of endodermal, ectodermal, and mesodermal origin (Ladd 2007). The endoderm gives rise to the digestive and respiratory tract, the ectoderm gives rise to skin and nervous tissue, and the mesoderm becomes the internal organ systems, including the heart. ESC lines develop from an undifferentiated stage resembling cells of the early embryo into terminally differentiated stages of the cardiogenic (Wobus et al. 1991, 1997; Maltsev et al. 1993, 1994; Miller-Hance et al. 1993), myogenic (Miller-Hance et al. 1993; Rohwedel et al. 1994; Rose et al. 1994), neurogenic (Bain et al. 1995; Fraichard et al. 1995; Strubing et al. 1995; Okabe et al. 1996), hematopoietic (Wiles and Keller 1991; Hole and Smith 1994; Keller 1995), adipogenic (Dani et al. 1997), or chondrogenic (Kramer et al. 2000) lineage, as well as into epithelial (Bagutti et al. 1996), endothelial (Risau et al. 1988), and vascular smooth muscle (VSM) cells (Risau et al. 1988; Weitzer et al. 1995; Drab et al. 1997). If the ESCs are exposed to appropriate cues, the developmental potential of ESC becomes more restricted to certain lineages as embryogenesis progresses and finally terminally differentiated to form various tissues. The in vitro cardiac differentiation of ES cells allows investigators (1) to understand developmental processes during the differentiation of stem cells into specialized cardiac cell types and early processes of commitment to specific cardiac lineages; (2) to study the effects of differentiation factors or xenobiotics on embryogenesis of cardiac tissue in vitro; and (3) to investigate pharmacological effects on functionally active cardiomyocytes (which are otherwise not available from a cell line).

Ultimately, the self-renewal ability of ESCs represent an unlimited ex vivo cell source for cardiac regenerative therapy (Thomson et al. 1998; Moore et al. 2005; Mummery et al. 2007; Siu et al. 2007a). In fact, genuine CMs with cardiac-specific structural and functional properties (Kehat et al. 2001) can be consistently differentiated from ESCs with various methods. Furthermore, various subtypes of CMs including pacemaker, atrial, and ventricular cardiomyocytes have been identified

(He et al. 2003). Despite recent advances in ESC-based cardiac regeneration, there are many obstacles to the clinical use of ESCs for cardiac regeneration. First, the efficiency of in vitro cardiac differentiation from ESC remains low (typically less than 1%), thus making it very difficult to achieve the number of CMs needed for therapeutic applications (Siu et al. 2007a). More importantly, the problem of immune rejection of hESC cell transplantation has limited clinical application. Last, ESC-derived cardiomyocytes display structurally and functionally immature phenotypes relative to their adult counterparts (Siu et al.2007a). The cells appear to have immature electrophysiological properties as well as underdeveloped $Ca^{2+}$ handling machinery (Fu et al. 2006; Au et al. 2009; Lieu et al. 2009), which not only results in ineffective contractile force generation but may also lead to creation of arrhythmogenic substrates, raising potential safety concerns for ESC-based cardiac therapy (Siu et al. 2007a). The following tables summarize the pros and cons of using different pluripotent cell sources for cardiac therapy.

## Human Induced Pluripotent Stem Cell (hiPSC)–Derived Cardiomyocytes as a Source for Cardiac Regeneration

A continuous supply of human cardiomyocytes is highly desirable, not only for screening drugs specifically targeting cardiac cells but also for cardiac toxicity testing for other, noncardiac drugs. In addition, the production of human cardiac cells bearing specific gene mutations would facilitate the development of improved in vitro methods for testing subsets of disease-related drugs and disease pathology. In the long term, cardiac regeneration may become an important part of cardiac interventional strategies as a means of supplementing or restoring contractile force in the failing heart. Because human embryonic stem cells are pluripotent and have the capacity for indefinite self-renewal, they are considered a potentially promising source of cardiomyocytes.

In recent years, it has been suggested that a novel potential source for cardiac cell regeneration is cardiomyocytes derived from human induced pluripotent stem cells (hiPSCs). hiPSCs can be generated from adult human dermal fibroblasts by transduction of a defined set of transcription factors to reprogram them back to an equivalent of the early embryonic state (Takahashi et al. 2007; Yamanaka 2007; Yu et al. 2007; Park et al. 2008). hiPSCs, resembling human embryonic stem cells (hESCs), possess remarkable self-renewal capacity and unquestioned potential to differentiate into three germ layers with mesoderm that will give rise to genuine cardiomyocytes (Zhang et al. 2009; Zwi et al. 2009). Being genetically identical to cells of the donor patients, patient-specific iPSCs avoid potential immune rejection and do not have the ethical issues specific to hESCs (Yamanaka 2007), thereby representing an attractive cell source for future cardiac regenerative therapy (Siu et al. 2007a).

However, the prerequisite to the ultimate clinical application is that hiPSC-derived cardiac derivatives display normal physiological characteristics. Transplantation of immature cardiomyocytes not only results in poor graft–host integration but may also lead to potential lethal arrhythmia (Zhang et al. 2002; Liao et al. 2010). Thereby, functional characterization of the stem cell-derived cardiomyocytes represents the crucial first step toward ultimate clinical application. Although electrophysiological properties such as ion channel profiling of hiPSC-derived cardiomyocytes have been reported (Gai et al. 2009; Yokoo et al. 2009; Zhang et al. 2009; Zwi et al. 2009), which are similar to those of hESC-derived cardiomyocytes and early fetal cardiomyocytes (Dolnikov et al. 2005, 2006; Fu et al. 2006; Siu et al. 2007a), there is a paucity of data concerning the calcium handling properties of hiPSC-derived cardiomyocytes, the key process underlying the excitation–contraction coupling (EC coupling). In this section, we discuss recent work that compared maturity of iPS-CMCs and hESC-CMCs in terms of calcium homeostasis, which increases our understanding of the development of excitability and EC coupling in the differentiating cardiomyocytes. To a greater extent, the establishment of an hiPSC-derived cardiomyocyte model may provide an opportunity to study the maturation process from the early developmental process of the cardiac tissue patient specifically. However, before confirming that this is comparable to a real situation, the iPSC-CMC model has to be investigated to see if the progress of maturation in iPSC-CMC is consistent with the hESC-CMC system.

To establish hiPSC models for studying cardiac phenotypes, appropriate differentiation strategies must be developed. Because the aggregates of pluripotent stem cells form an outer layer of (extra)embryonic endoderm, they have been termed embryoid bodies (EBs). The endoderm may serve as a crucial paracrine differentiation signal in EBs, as it is known that, in normal embryonic development, the endoderm is essential for signals to the anterior mesoderm during heart formation. However, the "spontaneous" cardiogenesis of undifferentiated stem cells is generally very inefficient, with only 1–2% of the cells within a beating EB or cell aggregate being cardiomyocytes. Much of the literature on improving these efficiencies includes activating specific developmentally relevant signaling pathways while the cells are growing as EBs. The Wnt and bone morphogenetic protein (BMP) signaling pathways, for example, have proved most potent in the mouse, although addition of ascorbic acid, serum-free conditions, low oxygen tension, and a variety of modifications to the growth medium or ways of forming cell aggregates have all been reported to enhance efficiency.

A prerequisite for using the pluripotent stem cells already mentioned is the efficient production of large and preferably homogeneous cell populations of cardiomyocytes. The following are the three major methodologies commonly used for cardiac differentiation of either hESC or hiPSC.

## Spontaneous Cardiac Differentiation

Formation of embryoid bodies (EBs) by suspension of stem cell colony pieces in a nonadherent dish will generate an embryo-like structure with three germ layers that will differentiate into all the varieties of cells present in our bodies. To induce

Feeder-free culture

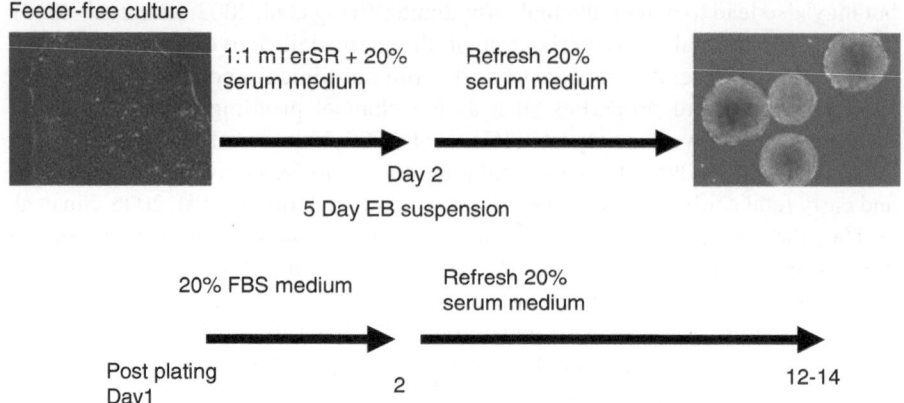

**Fig. 1.2** Protocol outline of spontaneous cardiac differentiation of pluripotent stem cells

differentiation, hiPS cells were dispersed into small clumps with collagenase IV (1 mg/ml at 37 °C for 20 min; Life Technologies, Carlsbad, CA, USA). The cells were then transferred to plastic Petri dishes, where they aggregated to form embryoid bodies (EBs) and were cultured in suspension for 10 days. The EBs were then plated on 0.1% gelatin-coated culture dishes in 20% fetal bovine serum (FBS) culture medium and examined daily for the appearance of spontaneous contractions (Zwi et al. 2009). The foregoing method is easy to carry out and efficient, but the yield of cardiomyocytes would be generally too low for downstream molecular studies, with only 1–2% of cardiomyocytes found in a beating EB. The other drawback is the overgrowth of fibroblasts in the culture as induced by a high percentage of FBS. Batch-to-batch variation of serum may also account for the unstable yield of cardiogenesis. It is highly recommended to perform a quality test for each lot of serum for the corresponding cardiac differentiation potential (Fig. 1.2).

## Mesoderm Enrichment by Wnt-Signaling Growth Factors

Undifferentiated hiPSCs were cultured on Matrigel (BD Biosciences, Bedford, MA, USA)-coated dishes with mTeSR medium (Stem Cell Technologies, Vancouver, BC, Canada). Cardiac differentiation was performed according to modification of a published protocol (Yang et al. 2008). Briefly, the cells were dissociated into clumps using 1 mg/ml Dispase (Invitrogen, Carlsbad, CA, USA) and cultured in suspension using low-attachment plates to form embryoid bodies (EBs). The embryoid bodies were transferred onto gelatin-coated plates in StemPro-34 basal medium (Invitrogen) supplemented with 2 mM glutamine (Gibco, Carlsbad, CA, USA), 4 mM monothioglycerol (Sigma-Aldrich, St. Louis, MO, USA), 50 µg/ml ascorbic acid (Sigma-Aldrich), and 0.5 ng/ml BMP-4

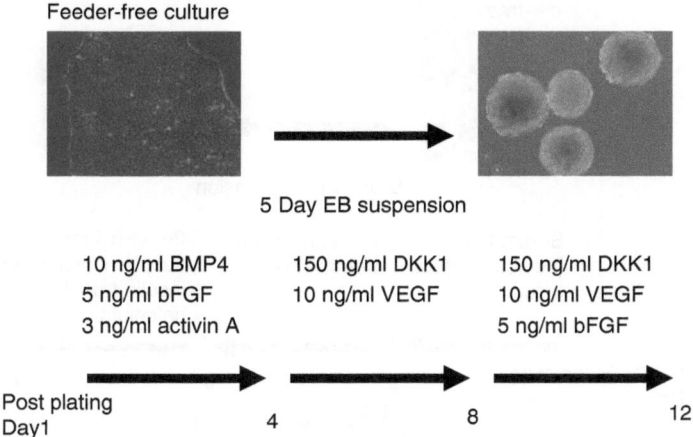

**Fig. 1.3** Protocol outline of cardiac differentiation of pluripotent stem cells by Wnt-signaling manipulation

(R&D). In addition, the medium was further supplemented with various cytokines according to the following sequence: post-plating days 1–4, 10 ng/ml BMP-4, 5 ng/ml basic fibroblast growth factor (bFGF), and 3 ng/ml activin A (Stemgent, San Diego, CA, USA); post-plating days 4–8, 150 ng/ml DKK1 (Gibco, Karlsruhe, Germany) and 10 ng/ml vascular endothelial growth factor (VEGF); and after day 8, 150 ng/ml DKK1, 10 ng/ml VEGF (Peprotech, Rocky Hill, NJ, USA), and 5 ng/ml bFGF (Peprotech). The growth factors were dissolved in Dulbecco's phosphate-buffered saline (DPBS) with 0.1% bovine serum albumin (BSA) at pH 5.5 (Fig. 1.3).

## Endoderm-Like Cell (END)-2 Co-culture Method

To induce a higher yield of cardiomyocytes, undifferentiated iPSCs were co-cultured with mouse visceral endoderm-like cells (END-2) as previously described (Mummery et al. 2003). The endoderm plays an important role in the differentiation of cardiogenic precursor cells that are present in the adjacent mesoderm in vivo. Briefly, undifferentiated iPSCs were dissociated into clumps using dispase (Invitrogen) and cultured in suspension using low-attachment plates to form embryoid bodies (EBs). The EBs were transferred onto irradiated END-2 cells for co-culture in serum-free condition for 9 days. Beating outgrowths from iPSC EBs were microsurgically dissected 20 days after induction of cardiac differentiation (Fig. 1.4).

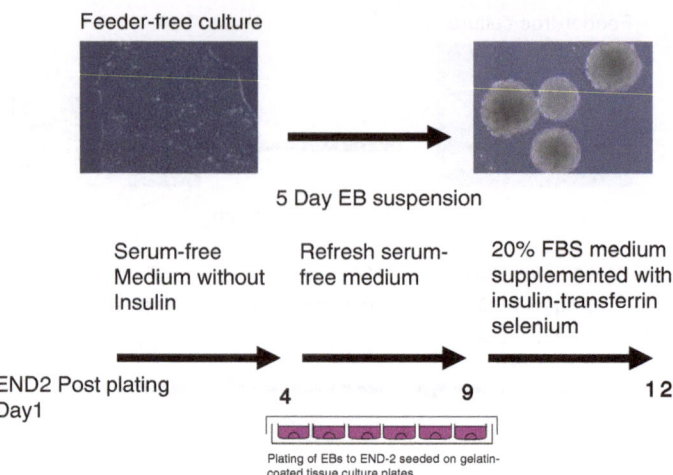

Feeder-free culture

5 Day EB suspension

Serum-free          Refresh serum-      20% FBS medium
Medium without      free medium         supplemented with
Insulin                                 insulin-transferrin
                                        selenium

END2 Post plating
Day1                        4            9            1 2

Plating of EBs to END-2 seeded on gelatin-
coated tissue culture plates

**Fig. 1.4** Cardiac differentiation of pluripotent stem cells by END-2 co-culture

## Calcium Homeostasis in Cardiomyocytes

Calcium homeostasis governs contractility of cardiomyocytes and is implicated in cardiomyopathy. In fact, dynamic alteration of cytosolic calcium concentration plays the central role in cardiac excitation–contraction coupling. Any abnormality in calcium release during the systolic stage and calcium reuptake during the diastolic phase will lead to heart pumping defeats. The sarcoplasmic reticulum (SR) functions as a storage vesicle for buffering to prevent so-called calcium leakage from the intracellular to extracellular region.

In adult cardiomyocytes, a small calcium influx through L-type calcium channels during membrane depolarization in the systolic phase leads to the inward calcium movement that triggers contraction subsequently. The inward current accounts for phase 2 of the action potential plateau. However, this small influx of $Ca^{2+}$ is not sufficient to trigger a large calcium release, and thus the signal is amplified by additional ryanodine receptor (RyR)–mediated release from the internal calcium store at the SR (Itzhaki et al. 2006; Bers 2008). This step is known as calcium-induced calcium release (CICR), which is the primary mechanism linking electrical excitation and mechanical contraction in cardiomyocytes (Fabiato 1983; Bers and Perez-Reyes 1999; Bers 2002) (Fig. 1.5).

The $Ca^{2+}$-sensitive RyRs are located in the vicinity of L-type calcium channels on the transverse tubules (t-tubules) and are responsive to a local surge in transsarcolemmal $Ca^{2+}$ influx (Itzhaki et al. 2006). The $Ca^{2+}$ flux from the membranous L-type calcium channel from the exterior and SR release through RyRs remarkably increases $[Ca^{2+}]_i$ (Cheng et al. 1993; Niggli 1999) and facilitates $Ca^{2+}$ binding to the myofilament regulatory protein, troponin C, which causes switching of cardiomyocyte contractile fibers. During diastole, to allow relaxation, calcium is actively removed from the cytosol, mainly by reuptake through SR calcium-ATPase

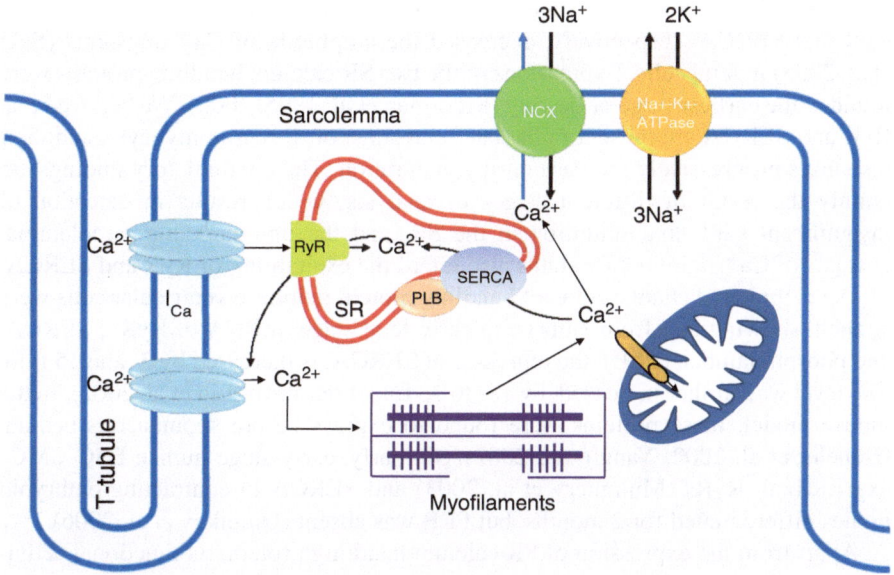

**Fig. 1.5** Mechanism of Ca$^{2+}$ homeostasis in ventricular myocytes (Bers 2002). *ICa*, L-type Ca$^{2+}$ channel; *NCX*, Na$^+$/Ca$^{2+}$ exchanger; *SR*, sarcoplasmic reticulum; *RyR*, ryanodine receptor; *SERCA*, SR calcium-ATPase; *PLB*, phospholamban

(SERCA) back into the SR and via sodium/calcium exchanger (NCX1) out of the cells (Bers 2008). The activities of these two systems are dependent on the Ca$^{2+}$ concentration gradient between sarcolemma and SR, the abundance of SERCA on SR and [Na$^+$]$_i$ for NCX, and also on membrane potential (Blaustein and Lederer 1999).

## Sarcoplasmic Reticulum (SR) Governs Maturity of Cardiomyocytes

One of the key aspects of the EC coupling machinery is Ca$^{2+}$-induced Ca$^{2+}$ release (CICR) from the SR. In human adult myocardium, the SR is the major intracellular Ca$^{2+}$ store, which accounts for approximately 70–90% of cytosolic Ca$^{2+}$ release and uptake for the contractile machinery (Bers and Despa 2006). However, in the fetal heart, the SR is structurally and functionally underdeveloped (Pegg and Michalak 1987). The SR vesicle has a lower volume, resulting in limited capacity for Ca$^{2+}$ buffering (Nakanishi et al. 1988). SR Ca$^{2+}$ is released via RyR2, as a central regulator of the CICR mechanism, and greatly contributes to EC coupling. The fetal cells are unresponsive to utilization of the RyR inhibitor ryanodine (Nakanishi et al. 1988; Liu et al. 2002), and it is therefore suggested that fetal cardiomyocytes are mostly dependent on transsarcolemmal Ca$^{2+}$ influx through the L-type calcium channel and efflux through NCX for contraction generation. In contrast to such findings, other reports challenged that ryanodine and thapsigargin, blockers of RyR and SERCA, respectively, decreased the amplitude of $Ca^{2+}$ transients (Seki et al. 2003) in fetal cells. Expressions of the two SR calcium handling proteins were found at the earliest stage of beating (Moorman et al. 1995, 2000). We believed that RyR and SERCA are present in the late fetal stage of the cardiomyocyte, and their responses progressively increase during maturation. The contradictory findings are mainly the result of different stages of analysis, which results in detection of insignificant $Ca^{2+}$ flux inhibition in the SR, and the unremarkable sarcolemmal changes of $Ca^{2+}$ in immature status also retard the excitability of RyR and SERCA.

Developmental changes of $Ca^{2+}$ handling protein in mouse ventricular cells were studied, showing that from embryo to early fetal stage, mRNA of RyR2, SERCA, and phospholamban (PLB), the repressor of SERCA, is increased by 3- and 15 fold. The level was further increased, by 18- to 33 fold, after birth (Liu et al. 2002). In the mouse model, these proteins were found to express before spontaneous beating (Boheler et al. 2002; Yang et al. 2002). Similarly, early-stage human ESC-CMCs express both RyRs (Mummery et al. 2003) and SERCA in contracting embryoid bodies differentiated for 2 months, but PLB was absent (Dolnikov et al. 2006).

Apart from the expression of SR calcium handling protein, its functional activities also determine the maturity of cardiomyocytes. Caffeine was used as a tool to determine the role of SR calcium in hESC-CM (Sauer et al. 2001; Satin et al. 2008). The mechanism is based on increased sensitivity of RyR2 to $Ca^{2+}$ and thereby releasing SR $Ca^{2+}$ (Rousseau and Meissner 1989). On the other hand, the hESC-CM SR calcium store is further confirmed by ryanodine inhibition on RyR-sensitive $Ca^{2+}$ release. The amplitude and frequency of $Ca^{2+}$ sparks and volume of caffeine-releasable SR store are reported to be increased time dependently during differentiation. RyR-sensitive $Ca^{2+}$ was also elevated with the developmental process (Sauer et al. 2001). Overall, there are three major criteria that govern the volume of SR $Ca^{2+}$: (1) increased Ca influx from the extracellular region to the interior through L-type $Ca^{2+}$ channels; (2) increased SR $Ca^{2+}$ uptake (depends on balance between SERCA and PLB); and (3) reduced $Ca^{2+}$ extrusion through NCX.

In our previous study, the responses of hiPSC-CMC to the RyR2 opener, caffeine, and the blocker, ryanodine, were much lower, implying there is an underdeveloped SR. However, deviated results were found in the study by Germanguz, who used H9 as the hESC-CMC group for comparison (Table 1.1) (Germanguz et al. 2009). The large differences may be caused by variation in differentiation stages among different cell lines; because HES3 and H7 were employed in our study, to some extent, the stage of maturation for investigation may not be consistent among different studies. Another reason may be deviation in the protocol of cardiac differentiation used, because the generated cardiomyocytes tend to show a fetal ventricular-like phenotype from various hESC lines (Mummery et al. 2003; Passier et al. 2005), while Wnt-signaling manipulation by adding growth factor BMP-4 and activin A shows an even distribution of different cell types. The cell picked for calcium imaging is required to be analyzed because ventricular, atrial, and nodal cells show their unique calcium homeostasis properties whereas atrial cells were reported to lack t-tubule structure (Shiels and White 2005). To improve

**Table 1.1** Comparison of the advantages and disadvantages of different types of cells for transplantation

| Cell source | Advantages | Disadvantages | References |
|---|---|---|---|
| Skeletal myoblasts | • Easy to obtain by isolation from skeletal muscle biopsies<br>• Possible for ex vivo expansion with *sufficient* quantities for autologous transplantation | • A lack of connexin-43 expression after in vitro differentiation resulting in failure of electrical integration with the host myocardium | Menasche et al. (2003), Roell et al. (2007) |
| Bone marrow (BM)–derived cells | • Possible for ex vivo expansion with sufficient quantities for autologous transplantation without immune rejection<br>• Paracrine effects of BM-EPCs and MSCs improve cardiac function | • Trans-differentiation into CMs may only be a rare event | Balsam et al. (2004), Murry et al. (2004), Nygren et al. (2004), Fedak (2008), Gnecchi et al. (2008) |
| Embryonic stem cells (ESCs) | • Able to be self-renewed in undifferentiated state without karyotypic alternation | • Efficiency of spontaneous in vitro cardiac differentiation is very low (<1%)<br>• Potential risk of immune rejection and teratoma formation<br>• Non-ethical with sacrifice of early embryos | Emanueli et al. (2005), Okita et al. (2007) |
| Induced pluripotent stem cells (iPSCs) | • Patient specific and feasible for autologous transplantation | • Low yield of cardiac differentiation<br>• Requires the use of viral vectors, which may lead to oncogenesis and genome abnormality | Okita et al. (2007) |

the accuracy of the studies, simultaneous patch-clamp recording and calcium florescence confocal imaging should be performed, such that the chamber-specific identities of cardiac cells can be distinguished before calcium kinetic analysis.

## SR Junctional Proteins Play a Role in Calcium Flux Between Cytosol and SR

The calcium signals that trigger contraction in cardiomyocytes depend on the SR $Ca^{2+}$ store. The rate of release of $Ca^{2+}$ from SR to cytosol in response to membrane depolarization when an action potential arrives is highly dependent on the communication between the voltage-gated L-type $Ca^{2+}$ channel on the surface membrane and RyR2 on the SR. These communications are tightly regulated by several key proteins localized in junction with SR and sarcolemma (junctional SR), triadin (Trdn) and junctin (Jun), which have an important role in forming a functional SR $Ca^{2+}$ release complex (Fig. 1.6). It was shown that these two proteins interact directly in the junctional SR membrane and stabilize a complex that anchors calsequestrin (Casq2) to the RyR2 (Zhang et al. 1997). Indeed, Casq2, with a high affinity to form a complex with $Ca^{2+}$, functions as a $Ca^{2+}$ store in the SR, whereas triadin and junctin structurally form a complex with Casq2 that facilitates release of SR $Ca^{2+}$ through RyR2 (Gyorke et al. 2009). Knocking down either junctin or triadin reduced $Ca^{2+}$ release induced by KCl depolarization by 20–25%, whereas silencing of both SR junctional proteins suppressed release by 35%, which suggested that they have independent roles in SR $Ca^{2+}$ regulation (Meissner et al. 2009). On the other hand, the absence of Casq2 renders hearts susceptible to premature spontaneous SR $Ca^{2+}$ release and triggered irregular heartbeats as a result of the loss of Casq2 buffering that prevents leakage of SR $Ca^{2+}$ (Knollmann 2009).

Several groups have demonstrated that the complex of triadin, junctin, and calsequestrin contributes to regulation of RyR2 by luminal $Ca^{2+}$ (Gyorke et al. 2004; Qin et al. 2008), while others have demonstrated luminal $Ca^{2+}$ dependence in purified RyR2 channels that lack Casq2 (Jones et al. 2008; Kong et al. 2008), suggesting that multiple mechanisms are likely to contribute to luminal regulation. The evidence from *Casq2* null mice, exhibiting a drastic reduction in both junctin and triadin, demonstrated that the relationship between SR $Ca^{2+}$ content and SR $Ca^{2+}$ leak is more severe, which supports the important role of Casq2 for luminal regulation. Even a modest reduction in Casq2 by 25% in heterozygous *Casq2* null mice resulted in steeper changes in SR $Ca^{2+}$ content (Blanco and Mercer 1998).

In our previous study, mRNA expression of RyR2 and SERCA2a was significantly lower in two cell lines of hiPSC-derived cardiomyocytes compared to the hESC counterpart (Lee et al. 2011a). In addition, among the three cardiac SR luminal auxiliary proteins, the expression of triadin and junctin was significantly lower in hiPSC-derived cardiomyocytes compared with hESC-derived cardiomyocytes,

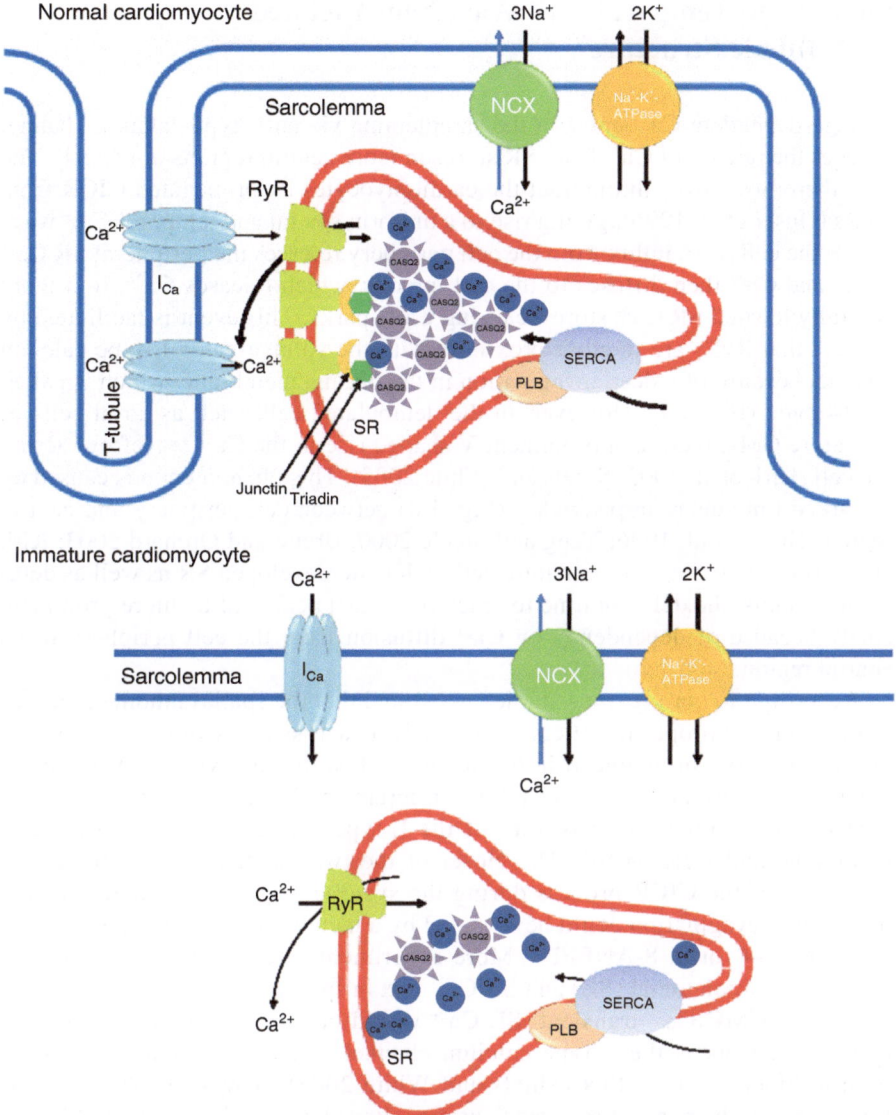

**Fig. 1.6** Significantly reduced expression of calsequestrin (CASQ2), Junctin (Jun), and Triadin (Trdn) in immature cardiomyocytes accounts for the reduced $Ca^{2+}$ release from and reuptake into SR. $I_{Ca}$, L-type $Ca^{2+}$ channel; *NCX*, $Na^+/Ca^{2+}$ exchanger; *SR*, sarcoplasmic reticulum; *RyR*, ryanodine receptor; *SERCA*, SR calcium-ATPase; *PLB*, phospholamban

whereas no significant differences in calsequestrin expression were observed between the two cell lines. Our findings were the first to point out reduced expression of calcium-pumping protein as well as three of the junctional proteins contributing to impaired SR function in hiPSC-CMCs.

# Spatial and Temporal Ca²⁺ Wavefront Dictated by T-Tubule Structure

Voltage-dependent $Ca^{2+}$ entry in the sarcolemma via an L-type calcium channel induces the release of $Ca^{2+}$ from SR stores in more central regions of the cell. The $Ca^{2+}$ therefore moves throughout the cardiomyocytes by propagated CICR from RyRs (Huser et al. 1996), giving rise to a uniform spatial and temporal $Ca^{2+}$ wave across the cell. $Ca^{2+}$ influx from the cell periphery releases the peripheral SR $Ca^{2+}$ store, and $Ca^{2+}$ then diffuses to the cell interior, which releases $Ca^{2+}$ from more centrally located SR $Ca^{2+}$ stores, causing $Ca^{2+}$ sparks. This event is facilitated by the fact that RyRs are in close proximity with the voltage-gated L-type calcium channel because of a deep invagination in the plasma membrane, which is called the t-tubule (Fig. 1.7). However, in the detubulated cell, such as atrial cells or immature CMs, there is a prominent V-shaped rise in the $Ca^{2+}$ wavefront across the cell (Kirk et al. 2003; Shiels and White 2005). This phenomenon is caused by a marked time delay in peak $Ca^{2+}$ (Fig. 1.8) between cell periphery and central region (Huser et al. 1996; Yang and Steele 2000; Brette and Orchard 2003; Kirk et al. 2003). In case of an immature cell with underdeveloped SR as well as detubulated status, the delay in time to reach the $[Ca^{2+}]$ peak will be more prominent solely because of dependency on $Ca^{2+}$ diffusion from the cell periphery to its central region.

According to our previous studies, we found that the spatial inhomogeneities in the temporal properties of calcium transients across the width of cardiomyocytes were more pronounced in hiPSC-derived cardiomyocytes than the hESC counterpart. Further studies in t-tubule structure and RyRs on structure are necessary to determine the proximity of the L-type calcium channel on the cell membrane and RyR on SR. The closer of the two candidates, the higher the efficiency of the CICR process during the systolic stage. The structure of the membrane invagination could be revealed by a membrane-bound voltage-sensitive dye, di-4- or di-8-ANEPPS. More experiments are required to investigate the role of sarcolemmal (SL) and SR $Ca^{2+}$ flux in spatial and temporal properties of the iPS-CMC $Ca^{2+}$ transient. SL $Ca^{2+}$ would be stimulated with (−)BAY K 8644, which opens the L-type calcium channel, whereas ryanodine could be used to block SR $Ca^{2+}$ flux (Shiels and White 2005). It was reported that the time to reach the peak of the transient was slowed across the whole cell by (−) BAY K 8644 application, reflecting the prolonged $Ca^{2+}$ entry through the L-type calcium channel (Shiels and White 2005). However, in the central region of the cell, SR inhibition by ryanodine slowed the upstroke velocity of the transient but not the absolute time to reach peak amplitude or the delay in the initiation of the rise in the $Ca^{2+}$ transient in the cell center. This observation suggested RyRs on the SR are dependent on diffusion of calcium from periphery to center.

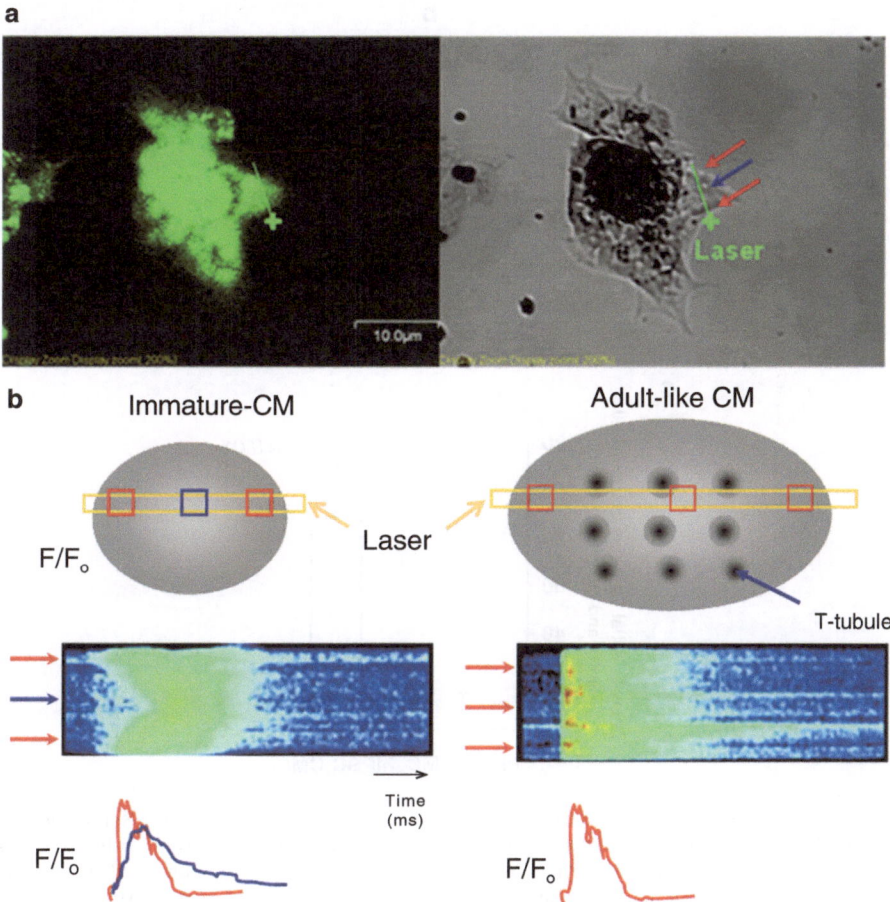

**Fig. 1.7** (**a**) Principle of calcium confocal line scan imaging to determine maturation of ESC-CM. Line scan (XT mode) was performed by drawing a line across the cell of interest for detection of temporal calcium fluorescence changes at a certain plane of the cell. (**b**) Principle of calcium confocal line-scan imaging to determine maturation of ESC-CM. In the immature cell lacking t-tubules, the Ca increase would not be synchronized. If we probe the boundary and the center of the cell, we will see that the Ca transient increase at the cell boundary would be faster with greater amplitude than that at the cell center. A U-shape calcium wave will be seen. For a successfully matured derived cell that has t-tubules, the Ca line-scan would show a synchronized Ca wave across the cell. There would be virtually no difference between the Ca transients at the center or the boundary of the cell (Lee et al. 2011a)

Apart from the crucial role of RyR in CICR, SR appears to play a role in Ca²⁺ resequestration (recycling) during relaxation because SR inhibition significantly slowed the transient directly (Bers 2002; Shiels and White 2005). The dependency of hiPSC-CMC in SR Ca²⁺ in the scope of spatial and temporal Ca²⁺ distribution remains to be investigated.

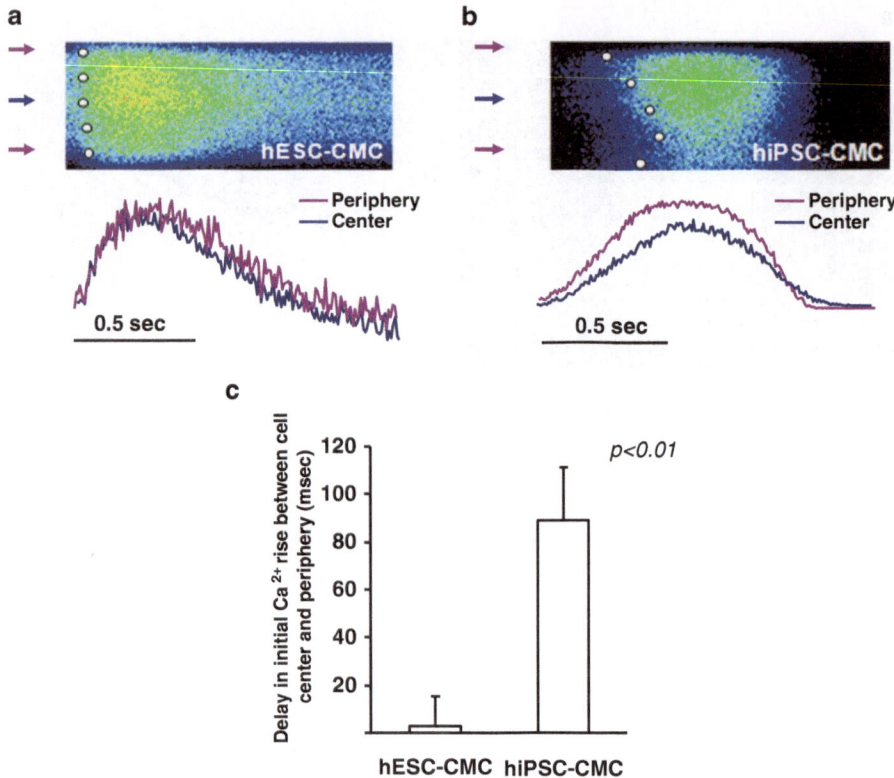

**Fig. 1.8** Transverse line-scan confocal of hiPSC- and hESC-derived cardiomyocytes. Representative tracings of temporal and spatial patterns of calcium transients of hESC-derived cardiomyocytes (**a**) and hiPSC-derived cardiomyocytes (**b**). (**c**) Time delay in the initial calcium rise between cell center and cell periphery in hESC- and hiPSC-derived cardiomyocytes (Lee et al. 2011a)

## IP3-Mediated Calcium Release Contributes to Whole-Cell Calcium Transients

The presence of inositol-1,4,5-trisphosphate receptors (IP3Rs) in hiPSC-CMCs demonstrated the important contribution of this alternative pathway to $Ca^{2+}$ handling in these cells (Itzhaki et al. 2011). IP3-dependent signaling has been shown to be crucial during the process of cardiac development, for which, in fact, in the embryo the IP3R is reported to be the first expressed $Ca^{2+}$ release channel (Rosemblit et al. 1999; Poindexter et al. 2001). However, in adult cardiomyocytes, IP3R contribution to cardiac physiology has remained controversial (Moschella and Marks 1993; Marks 2000; Bers 2002; Kapur and Banach 2007). In both mouse (Mery et al. 2005; Kapur and Banach 2007) and human (Satin et al. 2008; Sedan et al. 2008) ESC-CMCs, it has been demonstrated that an IP3-releasable $Ca^{2+}$ pool is expressed and functions, which accounts for their automaticity features. To evaluate the potential role of an IP3-releasable $Ca^{2+}$ pool in hiPSC-CMCs, immunocytostaining results revealed the presence of the IP3R that strongly distributed around the nucleus, resembling that observed in neonatal rat cardiomyocytes (Jaconi et al. 2000), mouse ESC-CMCs (Mery et al. 2005), and hESC-CMCs (Satin et al. 2008).

To assess the potential contribution of IP3R activity to the whole-cell $[Ca^{2+}]_i$ transient modulation in hiPSC-CMCs, the effect of IP3R blockade was tested by utilizing two different antagonists. 2-Aminoethoxyphenyl borate (2-APB) application resulted in a significant decrease in whole-cell $[Ca^{2+}]_i$ transient amplitude and slowed whole-cell $[Ca^{2+}]_i$ transient frequency dose dependently. Consistent with these results, application of 2 μM U7312 [a phospholipase C (PLC) inhibitor] led to significantly diminished whole-cell $[Ca^{2+}]_i$ transient amplitude and a slowing of whole-cell $[Ca^{2+}]_i$ transient frequency. Blocking the activation of PLC inhibits a receptor-stimulated increase in the production of the second-messenger IP3 required as a trigger for IP3R-mediated $Ca^{2+}$ release (Kapur and Banach 2007; Sedan et al. 2008). These observations imply that an functional IP3-releasable $Ca^{2+}$ pool is present in hiPSC-CMCs and contributes to the modulation of $Ca^{2+}$ handling in these cells (Fig. 1.9). The current study described was limited to prove the function of

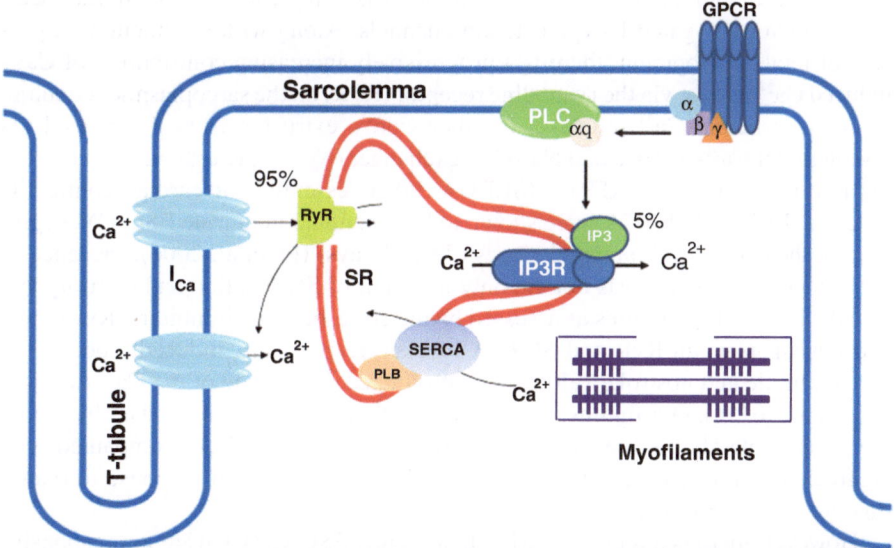

**Fig. 1.9** Contribution of IP$_3$-releasable calcium pool from SR in whole-cell calcium transient (estimated to be 5% in total). IP$_3$ is a soluble molecule and is capable of diffusing through the cytoplasm to the ER, or the sarcoplasmic reticulum (SR), once produced by phospholipase C (PLC). IP$_3$ is able to bind to the IP3R receptor on a ligand-gated $Ca^{2+}$ channel that is present on the surface of the SR. The binding of IP3 to IP3R triggers the opening of the $Ca^{2+}$ channel and the release of $Ca^{2+}$ into the cytoplasm (Moschella and Marks 1993). In heart muscle cells, this increase in $Ca^{2+}$ activates the ryanodine receptor-operated channel on the SR and results in further increases in $Ca^{2+}$ through a process known as calcium-induced calcium release. IP$_3$ may also activate $Ca^{2+}$ channels on the cell membrane indirectly by increasing the intracellular $Ca^{2+}$ concentration (Bers 2008). *GPCR*, G protein-coupled receptor

IP3-induced calcium release in iPSC-CMCs, but no comparison with human ESC-CMCs or fetal CMCs was made. Further study is required to evaluate the extent of contribution with IP3-$[Ca^{2+}]_i$ release in the two cell types.

## Calcium Handling Properties of hES-Derived Cardiomyocytes

Apart from the quantity of cardiomyocytes yield required to be investigated, the quality of ES cardiomyocytes needs to be assessed by functional characterization. To avoid arrhythmic risk during cell therapy, the functional properties of ESC-CMCs need to match with the host adult cardiomyocytes, which includes elicitation–contraction (EC) coupling, in terms of calcium homeostasis, ion channel electrophysiology, and contractile strength. In our previous studies, we aimed to examine the functional phenotype of ESC-CMC by calcium handling because calcium homeostasis is the earliest event happened in the initial stage of beating cardiomyocytes, and SR $Ca^{2+}$ store, a crucial component in governing contraction, is well known to indicate the maturity of cardiomyocytes.

Increase of intracellular $Ca^{2+}$ concentration is caused by transmembrane $Ca^{2+}$ influx via voltage-gated L-type calcium channels. Along with the maturation process of fetal development, there is a progressively increasing contribution of $Ca^{2+}$-induced $Ca^{2+}$ release via the ryanodine receptor (RyR) on the sarcoplasmic reticulum (SR). An efficient $Ca^{2+}$ reuptake, in other words, extrusion from the cytosol via sarcolemmal $Ca^{2+}$-ATPase and $Na^+$-$Ca^{2+}$ exchanger (NCX) and especially via sarcoplasmic reticulum $Ca^{2+}$-ATPase (SERCA 2a), is essential for cardiac contractile function. RyR and SERCA were shown to be functional in murine ESC-CM beginning in the early developmental stages $(7 + 2–4$ days) (Fu et al. 2006). Presence of the crucial SR proteins was not the only indicator of SR function and maturity, but also the functional activities as assessed by the corresponding inhibitors. Ryanodine and thapsigargin, an RyR and SERCA blocker, respectively, induced a decline of the amplitude and upstroke velocity and prolonged decay time of the $Ca^{2+}$ transient (Lee et al. 2011a). Our findings agreed with Satin's studies mentioned earlier about the remarkable abundance of RyR2 on H9.2 hESC-CMC SR, accompanied with significant caffeine-releasable $Ca^{2+}$ store and a time-dependent response to ryanodine inhibition (Satin et al. 2008).

However, there is a report showing that human ESC-CMCs were not responsive to the blockers, which implied that their contractions are dependent on transmembrane $Ca^{2+}$ influx, but not SR. Liu's study revealed that a mixture of CMCs appeared in a differentiated hESC-CMC population with caffeine-sensitive cells that are responsive to ryanodine inhibition of RyR2 while the others were not (Liu et al. 2007). Deviation in results may be caused by differences in the perfusion setup of caffeine or ryanodine. Variations in cell preparation may also affect the results, because only single cells and small monolayered clusters promise direct contact of most of the cell surface with the applied caffeine.

# Calcium Handling Properties of hiPSC-Derived Cardiomyocytes

With the advances of recent technologies, cardiomyocytes generated from human induced pluripotent stem cells (hiPSCs) are suggested as the most promising candidate to replenish cardiomyocyte loss in regenerative medicine. Little is known about their calcium homeostasis, the key process underlying excitation–contraction coupling. In our previous studies, we investigated the calcium handling properties of hiPSC-derived cardiomyocytes and compared them with those from human embryonic stem cells (hESCs) (Lee et al. 2011a). We differentiated cardiomyocytes from hiPSCs (IMR90 and KS1) and hESCs (H7 and HES3) with established protocols. As a gold standard to ensure validity of comparing calcium homeostasis properties of cardiac cells derived from iPSCs and hESs, the percentages of nodal-like, atrial-like, and ventricular-like cardiomyocytes were not significantly different because the t-tubule system is absent in atrial cells, making the calcium homeostasis from cell exterior to intracellular drastically deviated from ventricular cells (Brette and Orchard 2003) (Fig. 1.10). Reverse-transcription polymerase chain reaction studies demonstrated the expressions of cardiac-specific markers, Nkx2.5 and $\alpha$- and $\beta$-myosin heavy chain (MHC), in both hiPSC- and hESC-derived cardiomyocytes. Staining of cardiac cytoskeletal proteins such as $\alpha$-actinin and tropomyosin further confirmed the identity of the cells.

In our recordings, although spontaneous rhythmic calcium transients could be readily observed in hiPSC-derived cardiomyocytes, these calcium transients were of even smaller amplitude, slower upstroke velocity, and slower reuptake compared with hESC-derived cardiomyocytes (Lee et al. 2011a). In addition, to determine whether the SR in hiPSC-derived cardiomyocytes is in fact functional, we investigated the effects of caffeine, a RyR2 opener, and ryanodine, a RyR2 blocker, on the calcium transients of hiPSC-derived cardiomyocytes (Fig. 1.11a, b). Better caffeine-induced calcium handling kinetics in hESC-CMs indicates a higher SR calcium store for buffering during EC coupling. Furthermore, in contrast with hESC-derived cardiomyocytes, ryanodine did not reduce the amplitudes, maximal upstroke, and decay velocity of calcium transients of hiPSC-derived cardiomyocytes. The lower responsiveness of the cells to the application of ryanodine indicated the immature function of SR relative to the hESC counterpart, which may be partly related to the lower expression levels of major calcium handling proteins including RyR in hiPSC-derived cardiomyocytes. Indeed, the lower amplitude of caffeine-induced calcium release accounts for the lower capacity of SR $Ca^{2+}$ store in iPS-CMs, which also partly contributes to the impaired calcium handling kinetics.

In addition, spatial inhomogeneity in temporal properties of calcium transients across the width of cardiomyocytes was more pronounced in hiPSC-derived cardiomyocytes than their hESC counterpart as revealed by line-scan calcium imaging. In adult cardiomyocytes, RyRs are brought in close proximity to L-type calcium channels in the t-tubular system, allowing calcium wavefronts to quickly penetrate to the interior of the cells. In stark contrast, ESC-derived cardiomyocytes exhibit spatial inhomogeneity in temporal properties of calcium transients across the width

**Fig. 1.10** Characterization of cardiomyocytes differentiated from H7-, HES3-hESCs and IMR90-, KS1-hiPSCs. The proportion of nodal-like, atrial-like, and ventricular-like cells in hESC-derived (**a**) and hiPSC-derived (**b**) cardiomyocytes was determined with action potential recordings with whole-cell patch-clamp (Lee et al. 2011a)

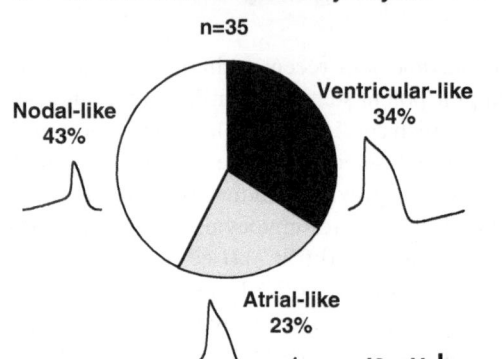

**a    hESC-derived cardiomy ocytes**

n=35

Nodal-like 43% · Ventricular-like 34% · Atrial-like 23%

40 mV / 400 ms

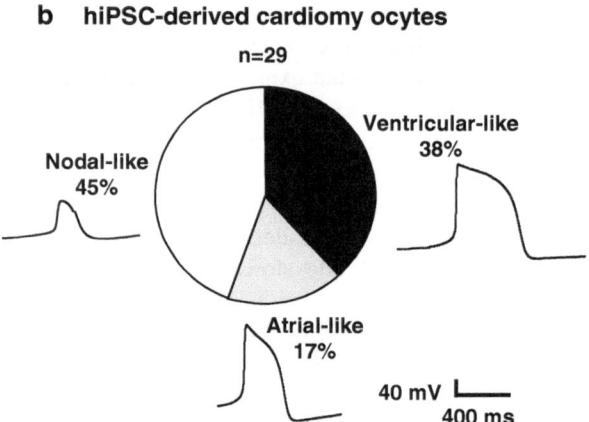

**b    hiPSC-derived cardiomy ocytes**

n=29

Nodal-like 45% · Ventricular-like 38% · Atrial-like 17%

40 mV / 400 ms

of the cells (Au et al. 2009; Lieu et al. 2009). Specifically, calcium wavefronts initiate earlier and reach their peak faster at the periphery of the ESC-derived cardiomyocytes than at the center. As a result, the calcium wavefronts across the width of the cells as revealed by transverse line-scan images typically show a U-shaped rise with a marked delay in peak between the cell periphery and the cell center (Au et al. 2009; Lieu et al. 2009). Recently, this inhomogeneity in the calcium release wavefront that reduces the efficacy of the calcium-induced excitation–contraction coupling has also been demonstrated in adult cardiomyocytes from a failing heart (He et al. 2001; Louch et al. 2004; Song et al. 2006). In the present study, although both hiPSC- and hESC-derived cardiomyocytes exhibited this typical U-shaped rise in calcium wavefronts across the width of the cells, the delay between the cell periphery and the cell center was significantly longer in hiPSC-derived cardiomyocytes than the hESC counterpart. It has been suggested that this poor coupling between the calcium influx through L-type calcium channels and the calcium release from the SR through RyRs is secondary to the spatial separation of these

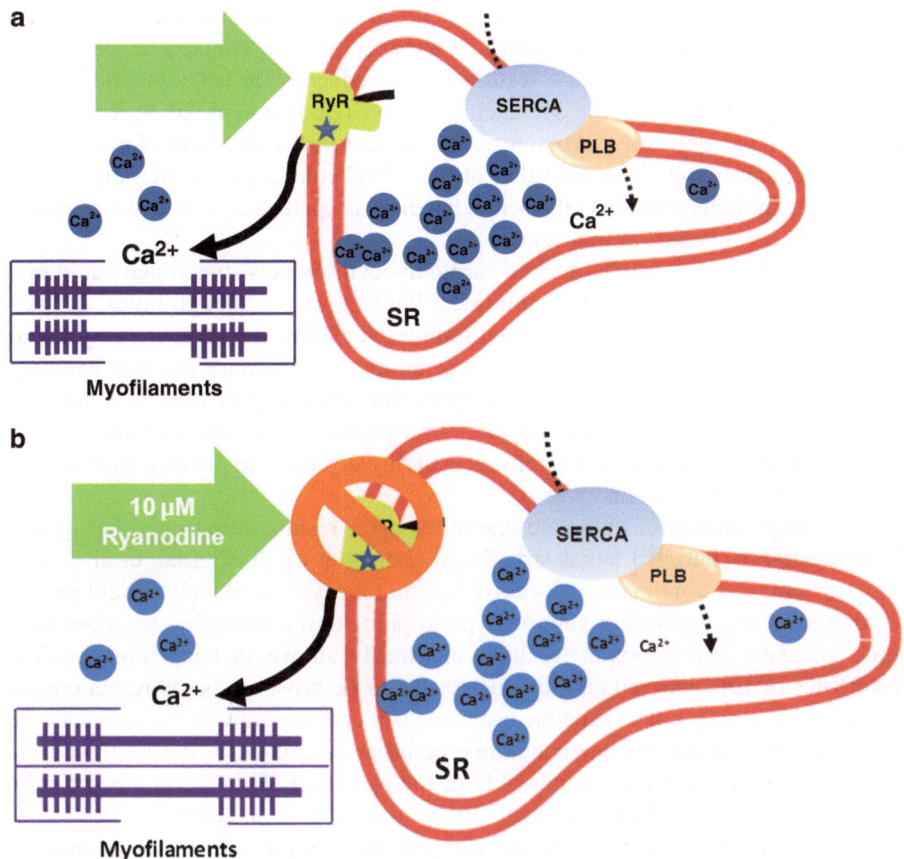

**Fig. 1.11** Mechanism of caffeine-induced (**a**) and ryanodine-suppressed (**b**) $Ca^{2+}$ homeostasis in ventricular myocytes. *SR*, sarcoplasmic reticulum; *RyR*, ryanodine receptor; *SERCA*, SR calcium-ATPase; *PLB*, phospholamban

ultrastructural organizations because of the lack of t-tubules (He et al. 2001; Louch et al. 2004; Song et al. 2006; Au et al. 2009; Lieu et al. 2009).

Furthermore, expressions of the key calcium handling proteins, including ryanodine recptor-2 (RyR2), sarcoplasmic reticulum calcium-ATPase (SERCA), junctin (Jun), and triadin (TRDN), were significantly lower in hiPSCs than in hESCs. The results indicated the calcium handling properties of hiPSC-derived cardiomyocytes are relatively immature compared to hESC counterparts. At the SR, triadin, a 32-kDa protein, together with a set of interrelated and interacting proteins including junctin and calsequestrin, forms a complex with RyRs and the juxtaposed L-type calcium channels in the sarcolemma. In addition to serving as a physical linker between RyRs and L-type calcium channels, triadin has also been shown to regulate calcium release via modulating the gating function of RyRs (Terentyev et al. 2005), as well as modifying the SR structure (Marty et al. 2009). In fact, in triadin-null hearts, the ablation of triadin results in significant reduction of junctional SR proteins, including RyR2, calsequestrin, and junctin, and a 50% reduction in the contacts between junctional SR and t-tubules, which translate into reduced SR calcium release and impaired negative feedback of SR calcium release (Chopra et al. 2009; Knollmann 2009). In the present study, the expression of triadin was significantly lower in hiPSC-derived cardiomyocytes than that of the hESC counterpart, which may partly explain the immature whole-cell calcium handling properties as well as the temporal inhomogeneity in calcium transient.

Taken collectively, hiPSC-derived cardiomyocytes possess functional but immature SR. This finding has potential clinical implications. Although hiPSC-derived cardiomyocytes, avoiding the potential immune rejection and ethic issues peculiar to hESC, hold promise for cardiac regeneration, the immature calcium handling properties of hiPSC-derived cardiomyocytes can result in poor functional integration and/or lethal arrhythmia after cellular transplantation, thus limiting their potential therapeutic applications. Thereby, it is highly desirable to develop strategies to drive functional maturation ex vivo.

As transplantation of immature cardiomyocytes results in not only poor graft–host integration but also lethal arrhythmia (Zhang et al. 2002; Liao et al. 2010), in vitro manipulation of stem cell-derived cardiomyocytes is needed to yield mature cardiomyocytes. Exploiting various hypertrophic stimuli including triiodonthyronine, ouabain, and hypoxia, we have previously shown that calcium handling properties of ESC-derived cardiomyocytes could be driven to be more mature (or adult-like) for potential therapeutic use, and such changes were associated with the upregulation of major calcium handling proteins as in ESC-derived cardiomyocytes by thyroid hormone, hypoxia-inducible factor 1-alpha (HIF-1$\alpha$) expression, or the cardiotonic steroid ouabain (Lee et al. 2010, 2011b; Ng et al. 2010).

First, according to our experience with murine ESC, $T_3$ supplementation upon cardiac differentiation favorably altered the calcium handling properties of ESC-derived cardiomyocytes, including larger calcium transients and faster rate of rise and decay of calcium oscillation. In addition, these cardiomyocytes also appeared to have a larger internal store of calcium as evidenced by the larger amplitude of caffeine-mediated calcium release. These phenotypic changes were likely associated with transcriptional upregulation of calcium handling proteins (RyR2, NCX-1, and SERCA-2a) (Lee et al. 2010). The underlying mechanism is shown in Fig. 1.12.

Second, in addition to cardiac differentiation, increased HIF-1$\alpha$ expression also drives the maturation of the ESC-derived cardiomyocytes, as demonstrated by the improved $Ca^{2+}$ handling and SR function in the transduced group (Ng et al. 2010). As indicated from the tracing of the spontaneous $Ca^{2+}$ transient, the amplitude of the $Ca^{2+}$ responses in cardiomyocytes derived from the HIF-1$\alpha$-transduced cells was significantly higher than that derived from the wild-type cells. The magnitude of alteration in caffeine-induced amplitude recorded from the HIF-1$\alpha$-transduced group compared with the WT group was similar to that of the spontaneous $Ca^{2+}$ transient, thus confirming the improved SR load, a sign of cardiomyocyte maturation. In addition to the amplitude, Ng and colleagues also observed an increase in the decay rate of the caffeine-induced $Ca^{2+}$ transient in the HIF-1$\alpha$-transduced

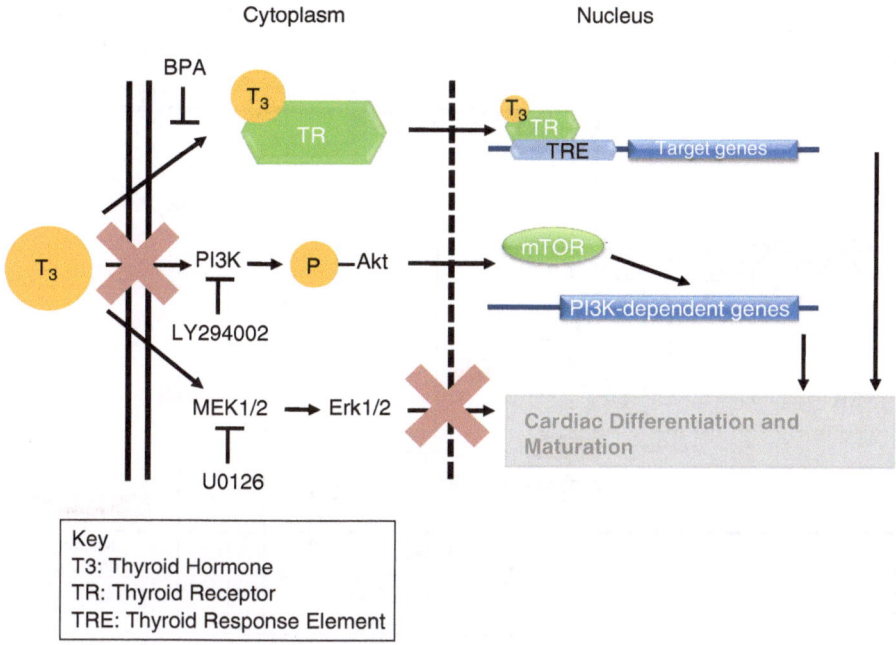

**Fig. 1.12** Summary of mechanistic study of $T_3$-mediated cardiac differentiation and maturation

group. Given that the RyR2 remains open in the presence of caffeine, such decrease in cytosolic $Ca^{2+}$ is more likely mediated by NCX1 rather than SERCA2a, even though mRNA levels of both SERCA2a and NCX1 were significantly higher in the HIF-1$\alpha$-transduced group. NCX1, rather than RyR2, appears to play the major role in mediating the excitation–contraction coupling in the middle stage of developing ESC-derived cardiomyocytes (Otsu et al. 2005).

Third, ouabain treatment favorably altered the calcium handling properties of mESC-derived cardiomyocytes, including larger calcium transients and a faster rate of rise and decay of calcium transients, and thus resulted in a stronger contractile force as well as excitation–contraction (EC) coupling, which electrically favors integration with the host tissue (Lee et al. 2011b). In addition, cardiomyocytes isolated from ouabain-treated EBs also appeared to have a larger internal store of calcium, as evidenced by larger amplitude of caffeine-mediated calcium release. These changes could be related to the corresponding upregulation of key calcium handling proteins in cardiomyocytes isolated from ouabain-treated EBs. Specifically, the upregulated ryanodine receptor could result in a faster rate of calcium release, while the rate of calcium transient decay (calcium reuptake back to SR) corresponds to the higher expression of *SERCA* in ouabain-treated cardiomyocytes. The slightly reduced NCX-1 expression reduced efflux of $Ca^{2+}$ to the extracellular region, thus further enhancing the amount of $Ca^{2+}$ reuptake into SR as well as myofibril contraction. The enhanced intracellular calcium concentration resulting from

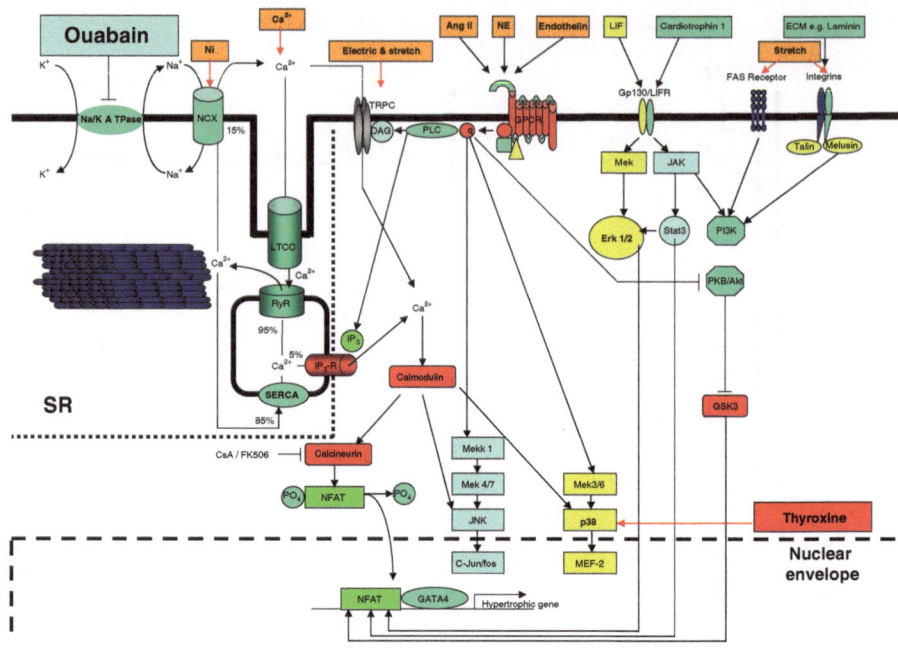

**Fig. 1.13** Summary of ouabain-induced signal transduction pathway in cardiomyocytes: the possible mechanism of ouabain-induced cardiac differentiation of mESC and maturation of mESC-derived CMC

$Na^+/K^+$-ATPase inhibition nonetheless remains another plausible mechanism for such improvement. The possible mechanism of ouabain-induced mESC-CMC maturation is summarized in Fig. 1.13.

Last, the problem of immaturity of hiPSC-CMCs might also be solved by the recent findings of Ng and colleagues, who reported cardiomyocytes derived from apoA-I-transduced IMR90-hiPSC cells exhibited improved calcium handling properties in both non-caffeine and caffeine-induced calcium transients, suggesting that apoA-I, a cholesterol transport-reversing candidate (Mineo and Shaul 2003; Mineo et al. 2006), plays a role in enhancing cardiac maturation (Ng et al. 2011). In comparison with the green fluorescent protein (GFP)-transduced control, exogenous apoA-I expression improves calcium homeostasis of human iPSC-derived cardiomyocytes, which exhibited more mature calcium handling properties including larger amplitude, higher maximal upstroke velocity, and maximal decay velocity of a spontaneous calcium transient.

To further assess the SR function, application of 10 μM ryanodine, a RyR blocker, substantially reduced the amplitude and upstroke values (approximately 60% reduction compared with drug-free recordings) of spontaneous calcium transients in cardiomyocytes derived from the LV-apoA-I transduced group. In contrast, cardiomyocytes derived from LV-GFP-transduced human iPSCs exhibited only a modest response to ryanodine (approximately 20% reduction in both amplitude and upstroke value of calcium transients), suggesting a more immature calcium handling apparatus. The mechanistic studies revealed that the mature $Ca^{2+}$ handling phenotype driven by ApopA1 secreted by the genetically overexpressed iPSC may be caused by accelerated cardiac differentiation or the direct effects of apoA-I on the cardiomyocytes. It was found that the promoted cardiac differentiation of ESCs and iPSCs used in the studies is related to the involvement of the BMP-4 signaling cascade, which is one of the downstream targets of ApopA1. The direct role of the BMP-4-mediated SMAD 1/5 activation pathway and the improved $Ca^{2+}$ homeostasis kinetics has yet to be determined. The possibility of driven maturation of hiPSC-derived cardiomyocytes with these strategies, as mentioned, requires further investigation (Table 1.2).

## iPSC-Derived Cardiomyocyte as a Potential Platform for Disease Modeling of the Impaired Calcium Handling–Related Syndrome

Initial proof-of-concept experiments in rodent MI models have demonstrated mESC-CMC disease modeling by manipulation of a calcium handling gene, for example, Junctin ablation in mouse ESCs to investigate its regulatory role in SR $Ca^{2+}$ homeostasis (Yuan et al. 2007). However, the pathogenic links between genotype and clinical phenotype are largely missing because of the lack of appropriate experimental models. The cellular and physiological differences between mouse and human, such as the large discrepancies in ion channel profile and multiple mutations in different exons of a gene, which might contribute to the subsequent phenotype, have made it increasingly important to develop more relevant human disease models for mechanistic studies. Regarding the differences in CMC calcium handling between human and rodent, the SR only contributes to 30% of $Ca^{2+}$ store whereas SR stores account for more than 70% in humans (Germanguz et al. 2009). Study of SR function would be more appropriate in hiPSC-CMCs rather than the rodent model.

In addition, ESC-CMCs show fetal-like electrical properties (e.g., spontaneous firing, phase-4 depolarization, prolonged action potentials) because of the different profiling of ion channel profiles in different stages of differentiation. As mentioned in the previous section, iPSC-CMC even showed a less mature phenotype than ESC-CMC. To ensure iPSC-CMC can be used for disease modeling, the timepoint at which mature phenotypes resembling adult cardiac tissue are acquired should be synchronized. We believed that the cardiomyocytes derived from iPSCs would be similar to the host tissue given a certain period of time, but the maturation process differs from hESC-CMCs. A parallel control of iPS-CMCs derived from a normal individual is required. It would be necessary to perform a functional assay proving that the electrophysiology of iPSC-CMC at the time of harvest has no difference from that of adult cells. Strategies developed for driven maturation of iPS-CMCs are desirable. According to our previous experience in mouse ESCs using a hypertrophic stimulus such as $T_3$ and ouabain, that enhanced volume of SR store as well as efficient calcium handing kinetics might be feasible to mature the cells (Lee et al. 2010, 2011b).

**Table 1.2** Summary of studies in the role of SR functions in calcium homeostasis properties of hESC- and hiPSC-derived cardiomyocytes

| Stem cell category | Cell line | Transient response to caffeine | Transient response to ryanodine | References |
|---|---|---|---|---|
| Human embryonic stem Cell-derived cardiomyocytes (hiPSC-CMC) | Hes2-, H1- | ~38% of the cells are responsive | Only caffeine-sensitive cells are responsive to ryanodine treatment | Liu et al. (2007) |
| | H9.2 Clone | +++ | +++ | Satin et al. (2008) |
| | H9.2 Clone | NA | – | Germanguz et al. (2009) |
| | Hes3, H7 | +++ | +++ | Lee et al. (2011a) |
| Human induced pluripotent stem cell-derived cardiomyocytes (hiPSC-CMC) | iPSC-clone1, 2 (generated from foreskin fibroblast) | + | ++ | Germanguz et al. (2009) |
| | IMR90-iPS, KS-1-iPS (generated from skin fibroblast) | + | – | Lee et al. (2011a) |

## Disease Modeling of Catecholaminergic Polymorphic Ventricular Tachycardia (CPVT)

Catecholaminergic polymorphic ventricular tachycardia (CPVT) is an inherited arrhythmia syndrome characterized by VT induced by adrenergic stress in the absence of structural heart disease and a high incidence of sudden cardiac death (Leenhardt et al. 1995; Laitinen et al. 2001; Cerrone et al. 2009; Hayashi et al. 2009). The patient who suffers form CPVT is normal in daily life, but ventricular tachyarrhythmias (VT) are commonly found during exercise or other vigorous adrenergic stimulation. The causative genes of CPVT are known as RyR2 and CASQ2, which encode the cardiac isoform of the $Ca^{2+}$ release channel, ryanodine receptor, and the $Ca^{2+}$-binding protein calsequestrin (Lahat et al. 2001; Laitinen et al. 2001; Cerrone et al. 2009). The mutations are commonly identified in approximately 60% of patients with CPVT (Watanabe and Knollmann 2011). The pathological consequences of such disease are related to the mutated or reduced level of calsequestrin or RyR2 that reduces the amount of bonded $Ca^{2+}$ ions in the SR, causing calcium leakage or destabilizing the RyR2 response to CICR, and results in sudden cardiac attack caused by arrhythmia related to delayed afterdepolarization (DAD) that shows a typical irregular pattern on the ECG cardiogram.

SR $Ca^{2+}$ release can occur in the absence of membrane depolarization through a mechanism referred to as spontaneous $Ca^{2+}$ release that is facilitated by the presence of SR $Ca^{2+}$ overload (Orchard et al. 1983; Marban et al. 1986; Fabiato 1992; Lakatta 1992). Spontaneous SR $Ca^{2+}$ release dependence on the size of the SR $Ca^{2+}$ store is designated as store overload-induced $Ca^{2+}$ release (SOICR) (Jiang et al. 2004, 2005) (Fig. 1.14). Adrenergic stimulation and other inotropic agents such as digitalis that cause a sudden surge in cellular $Ca^{2+}$ load because of suppressed $Na^+/K^+$-ATPase activity result in lowered activity of NCX, which reduces $Ca^{2+}$ extrusion to the cell exterior and facilitates SR $Ca^{2+}$ uptake. In addition, increased ventricular automaticity SR $Ca^{2+}$ overloading is also related to deficiency of junctin (Jun), a physical linker of SR $Ca^{2+}$ to RyR together with triadin (Yuan et al. 2007). The increased risk of lethal consequences in junctin-null mice may be caused by increased frequency of stress-induced DADs. The underlying rationale may be related to a less efficient supply of SR luminal $Ca^{2+}$ for CASQ and RyR binding, which is associated with its defective regulation.

The establishment of the iPSC-CMC-based platform facilitates our understanding of $Ca^{2+}$ regulation in the pathology of cardiac dysfunction. CPVT patients are carrying an inherited mutation in either Jun, CASQ2, or RyR2, or both the calcium handling proteins. With a pre-screening of gene mutation in patient-specific hiPSC-CMC, we are able to develop appropriate personalized strategies to improve the survival rate and life quality of CPVT patients. Possible approaches for suppressing CPVT are prevention of SR $Ca^{2+}$ overload and attenuating SOICR by RyR2 modulation (Fig. 1.15). As of today, the clinical management of CPVT relies on antiarrhythmic effects of beta-blockers and $Ca^{2+}$ channel blockers by reducing heart rate via suppressed $Ca^{2+}$ overload. However, effectiveness varies among individuals,

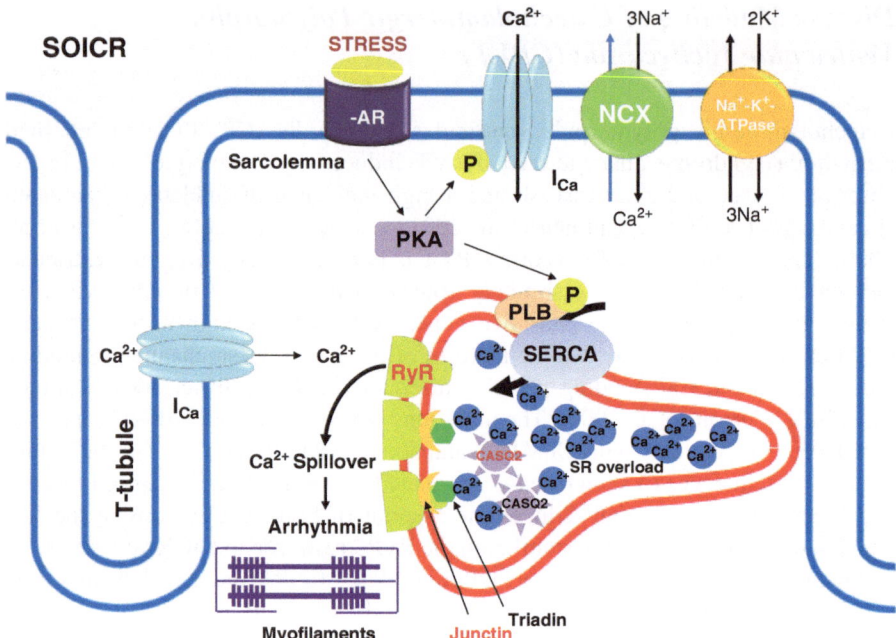

**Fig. 1.14** Store overload-induced $Ca^{2+}$ release (*SOICR*) and triggered arrhythmia under stress: the mechanism of SOICR, in which spontaneous SR $Ca^{2+}$ release or spillover caused by SR overload, such as stimulation of stress via the beta-adrenergic receptor (*b-AR*)/*PKA*/*PLB*-signaling pathway. The calcium handling proteins highlighted in *red* are the potential mutated candidates that play a role in SOICR. The mutated CASQ2 and RyR2 reduce the binding capacity of $Ca^{2+}$ in SR, while the juntin mutation will lead to loss of the physical link between RyR2 and SR $Ca^{2+}$, causing defects in strict regulation of $Ca^{2+}$ release through RyR

and recurrence of arrhythmia is commonly observed. Recently, in vitro data showed that the $Ca^{2+}$ channel blocker analogue flecainide, also a $Na^+$ channel blocker, inhibits RyR2 channel activity and thus suppresses SOICR (Hilliard et al. 2010). Other investigators also tested the efficacy of verapamil on treating CPVT caused by abnormalities in calcium handling (Cerrone et al. 2009). Further in vitro results showed that verapamil may exert a $Ca^{2+}$ modulation effect through binding to RyR2 and inhibiting its activity (Valdivia et al. 1990). The controversy about the survival benefits of that drug among different groups has yet to be resolved by developing an iPSC-CMC-based platform to determine personalized medication for CPVT patients carrying either the RyR2, CASQ2, or Jun mutation.

Until now, only heterologous expression systems and genetic mouse models have been used to study the cellular and molecular aspects of CPVT-related RYR2 mutations. Jung's team recently reported the generation of iPSCs from a CPVT patient carrying a novel missense RYR2 S406L mutation (Jung et al. 2011). This advance demonstrated the suitability of CPVT to recapitulate molecular and physiological aspects of the disease phenotype.

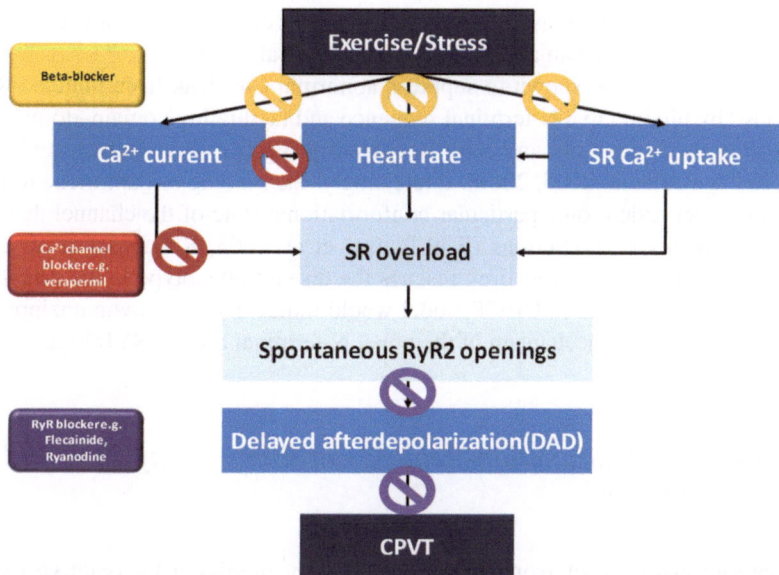

**Fig. 1.15** Mechanism-based approach for drug therapy in catecholaminergic polymorphic ventricular tachycardia (*CPVT*). Flecainide, beta-blocker, and Ca²⁺ channel antagonists independently target the underlying mechanisms responsible for CPVT. It was proposed that flecainide inhibits arrhythmias by RyR2 channel block and by Na⁺ channel block (Watanabe and Knollmann 2011)

Dantrolene, a hydantoin derivative that acts as a muscle relaxant, is currently the sole and most effective treatment for malignant hyperthermia, a rare life-threatening familial disorder caused by mutations in the skeletal ryanodine receptor (RYR1) (Kobayashi et al. 2009). Recently, dantrolene has been shown to target a corresponding sequence in RYR2 (Paul-Pletzer et al. 2005) and to improve intracellular Ca²⁺ handling in failing cardiomyocytes using a canine heart failure model (Kobayashi et al. 2009) and a murine arrhythmia model of CPVT1 (Kobayashi et al. 2010; Uchinoumi et al. 2010).

The establishment of the CPVT hiPSC disease model mainly revealed insights in two novel mechanisms. The first mechanism suggested that RYR2 mutants suppressed the binding affinity of the channel's auxiliary stabilizing protein FKBP12.6 and that this is further aggravated in situations of catecholamine-induced hyperphosphorylation of RYR2 with consequential dissociation of FKBP12.6 and Ca²⁺ leakage from the SR (Marx et al. 2000; Wehrens et al. 2003; Lehnart et al. 2008). There is also increasing evidence that clearly demonstrates that alterations in FKBP12.6-RYR2 interaction are unlikely to be the common cause of CPVT1 (George et al. 2003; Jiang et al. 2005; Liu et al. 2006; Xiao et al. 2007; Guo et al. 2010). Alternatively, it has been proposed that RyR2 mutations in the N-terminal and central regions of the protein weaken interactions between these two domains that are critical in stabilizing the closed state of the channel, resulting in an increased open probability and increased risk of $Ca^{2+}$ leakage during stress-induced SR $Ca^{2+}$ overload (Ikemoto and Yamamoto 2000; Tateishi et al. 2009).

Dantrolene has been shown to suppress abnormal $Ca^{2+}$ leak from mutated RYR1 and RYR2 by binding to a N-terminal sequence and facilitating domain–domain contacts within the N-terminal and central regulatory regions (Paul-Pletzer et al. 2002, 2005; Kobayashi et al. 2005, 2009). Interestingly, the binding of dantrolene to RyR2 seems to be dependent on a particular conformational state of the channel that takes place only in disease conditions (Paul-Pletzer et al. 2005; Kobayashi et al. 2009). Therefore, the results that this drug rescues the disease phenotype in the established patient-specific iPSC-based CPVT model would indicate that "domain unzipping" is likely to be the patho-mechanism of the novel N-terminal S406L-RYR2 mutation.

## Mitochondrial Dysfunction-Related Cardiomyopathy-Impaired SR $Ca^{2+}$ Reuptake

The pathophysiology of iron overload is clearly mediated by reactive oxygen species whereby the cytoplasmic labile iron pool becomes available for the conversion of cytotoxic $Fe^{2+}$ into $Fe^{3+}$. The process generates free radicals, including the highly reactive hydroxyl radical (Oudit et al. 2006). When there is a surge in iron ions, excessive free radical generation leads to increased peroxidation and damage to lipids, proteins, and nucleic acids, triggering cellular damage and depletion of antioxidants (Oudit et al. 2006). The effects of free radical production and oxidative stress in conditions of acute iron toxicosis and iron-overload cardiomyopathy have been well documented in patients with primary hemochromatosis, beta-thalassemia major, Fredriech's ataxia, and end-stage kidney disease (Young et al. 1994; Campuzano et al. 1996; Livrea et al. 1996; Lim et al. 2000; Walter et al. 2006).

Cardiac EC couplings are highly sensitive to changes in cellular redox state, leading to reduced systolic $Ca^{2+}$ because of impaired calcium release and elevated diastolic $Ca^{2+}$ resulting in impaired $Ca^{2+}$ reuptake back into SR, causing defeats in systolic and diastolic function characteristic of iron-overload cardiomyopathy (Oudit et al. 2004, 2006). The permeation of $Fe^{2+}$ through cardiac L-type $Ca^{2+}$ channels (LTCC) may be particularly relevant because this is the first step that delivers reactive ferrous iron from the exterior to the major regulators of EC coupling in cardiomyocytes. On the other hand, iron-induced mitochondrial dysfunction may also implicated in the altered energetic status of cardiomyocytes, which causes malfunction of $Ca^{2+}$ reuptake into SR via the energy-demanding SERCA or myofilament shortening (Ashrafian et al. 2011) (Fig. 1.16).

Apart from the sarcolemmal NCX, SR $Ca^{2+}$ handling proteins, and the mitochondrial $Ca^{2+}$ uniporter contributing to $Ca^{2+}$ cycling, sarcomeres also powerfully sequester $Ca^{2+}$ (Bers 2008). It is especially pertinent to hypertrophic cardiomyopathy (HCM) because, through increased calcium affinity, HCM-related sarcomeric mutations may further alter both the magnitude and pattern of the $Ca^{2+}$ transient by "calcium trapping," mediated by changes in the $Ca^{2+}$ affinity of troponin C, considered to be the major $Ca^{2+}$ buffer during $Ca^{2+}$ homeostasis. These abnormalities in EC

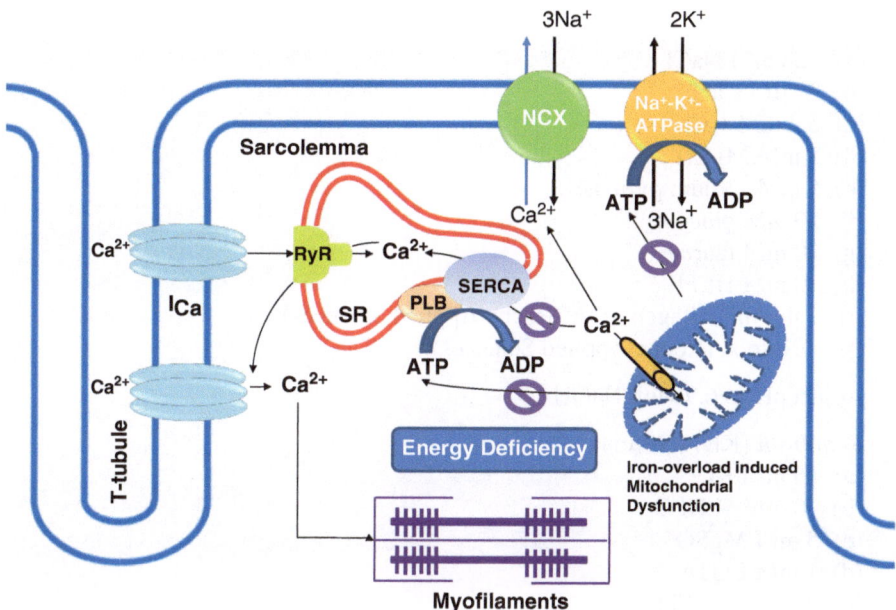

**Fig. 1.16** Iron-overload-induced mitochondrial dysfunction causing reduced myocardial energy generation

coupling caused by an iron-overloaded situation and HCM mutations are likely to contribute to cardiomyopathy and act as current therapeutic HCM targets; for example, inhibition of L-type $Ca^{2+}$ channels by diltiazem and neonatal gene transfer of Serca2a to relieve calcium trapping outside the SR (Ashrafian et al. 2011).

## Measurement of Cytosolic [Ca²⁺] by Fluorescence Confocal Microscopy

### *Materials*

1. Tyrode solution

    (a) 140 mM NaCl
    (b) 5 mM KCl
    (c) 0.4 mM $KH_2PO_4$
    (d) 1 mM $MgCl_2$
    (e) 1.8 mM $CaCl_2$
    (f) 10 mM glucose
    (g) 5 mM HEPES

    pH adjusted to 7.4 with NaOH

2. Dissociation buffer

  (a)  120 mM NaCl
  (b)  30 μM CaCl$_2$
  (c)  5.4 mM KCl
  (d)  5 mM MgSO$_4$
  (e)  5 mM sodium pyruvate
  (f)  20 mM glucose
  (g)  20 mM taurine
  (h)  10 mM HEPES
  (i)  1 mg/ml collagenase A (Roche Applied Sciences)
  (j)  DNase A (Roche Applied Sciences)

  Adjust pH to 6.9 with NaOH

3. *Kraftbrüh* (KB) solution
  (a)  85 mM KCl
  (b)  30 mM K$_2$HPO$_4$
  (c)  5 mM MgSO$_4$
  (d)  1 mM EGTA
  (e)  2 mM Na$_2$-ATP
  (f)  5 mM pyruvic acid
  (g)  5 mM creatine
  (h)  20 mM taurine
  (i)  20 mM D-glucose

4. Calcium calibration buffer kit (cat. no. C-3008MP; molecular probes)

  (a)  Zero free calcium buffer (component A)
  (b)  39 μM free calcium buffer (Component F)

## Calibration of Fluo-3

To quantify the exact amount of Ca$^{2+}$ present in the cell, a calibration curve of fluo-3 has to determine the linear range between calcium concentration and fluorescence intensity recorded before the measurement. In certain recording conditions, to verify the quality of Ca$^{2+}$ fluorescence dye, Fluo 3-AM, which would be metabolized by esterase in live cells to give fluorescence, fluo-3 must be calibrated with a fluorometer with a known calcium concentration ranging from 0 to 39 μM by reciprocal dilution of components A and F in the calibration buffer kit (Fig. 1.17). A linear relationship was shown at intracellular calcium concentrations up to 600 nM (Fig. 1.18). To maintain the accuracy of measuring trace amounts of Ca$^{2+}$ present in the buffer at standard concentration, use of a calibration kit is encouraged, according to a method described by Dr. Roger Tsien, to determine the dissociation constant ($K_d$) of the fluorescent Ca$^{2+}$ indicators at a chosen temperature, ionic strength, and pH ($K_d$ of Fluo-3 was determined at pH 7.4 at 37 °C in our experiment),

# Dissociation Constant (Kd) Calculator

Title:
Equation:                     $y = 1.02x + 0.362$
$R^2$:                            0.999
$K_d$ (µM):                    0.442

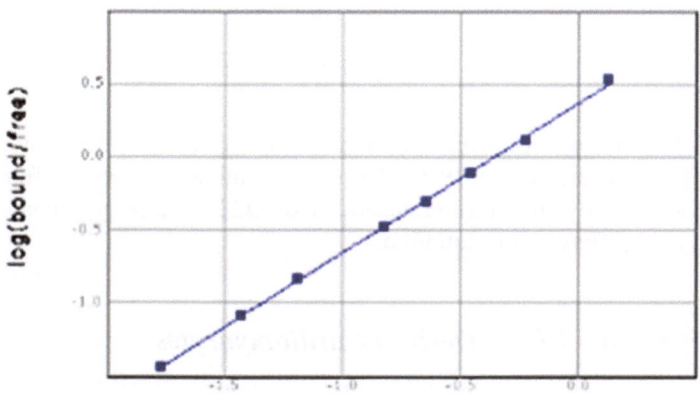

log([Ca2+]free)

Data Points

| [Ca²⁺]free (µM) | F | log([Ca²⁺]free) | log(bound/free)* |
|---|---|---|---|
| 0.0 | 751.0 | | |
| 0.017 | 1732.0 | -1.77 | -1.442 |
| 0.038 | 2903.0 | -1.42 | -1.082 |
| 0.065 | 4354.0 | -1.187 | -0.833 |
| 0.15 | 7669.0 | -0.824 | -0.487 |
| 0.225 | 9999.0 | -0.648 | -0.31 |
| 0.351 | 12890.0 | -0.455 | -0.12 |
| 0.602 | 16685.0 | -0.22 | 0.116 |
| 1.35 | 22458.0 | 0.13 | 0.528 |
| 39.0 | 28891.0 | | |

*bound/free = $(F-F_{min})/(F_{max}-F)$

**Fig. 1.17** $K_d$ determination of Fluo-3 Ca²⁺ fluorescence dye

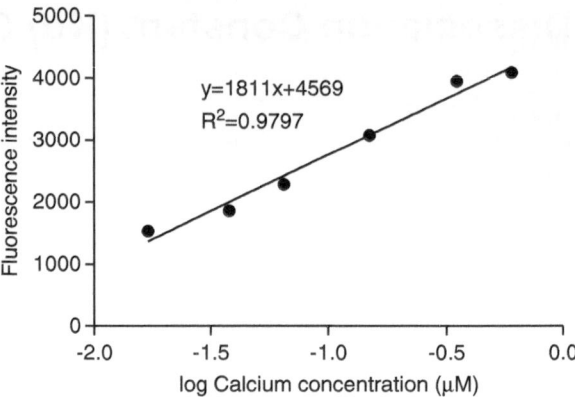

**Fig. 1.18** Calibration curve for Ca²⁺ concentration determined by Fluo-3 staining. Fluorescence was recorded by the confocal microscope for Ca²⁺ transient measurement. A calibration curve was plotted to determine the linear range for precise quantification

and such a protocol employs a reciprocal dilution method to minimize indicator concentration errors (Tsien and Pozzan 1989). The stock solutions with 39 μM free Ca²⁺ would be diluted by ethyleneglycoltetraacetic acid (EGTA) buffer to prepare buffers with free Ca²⁺ at different concentrations.

## *Isolation of hiPSC- and hESC-Derived Cardiomyocytes*

Differentiated hiPSC- and hESC-embryoid bodies containing beating outgrowths were microsurgically dissected using a glass knife on post-plating day 21 as previously described (Liu et al. 2007, 2009; Au et al. 2009; Lieu et al. 2009). Under calcium-free conditions, the beating cells were dissociated in medium containing 1 mg/ml collagenase B in presence of 60 U/ml DNase I (Roche Applied Sciences, Penzberg, Germany) at 37 °C for 30–45 min and were resuspended in Kraftbrüh (KB) solution at room temperature by shaking at 100 rpm for 1 h. The dissociated cells were then plated onto 0.1% gelatin-coated glass coverslips and maintained with 5% FBS differentiation culture medium for at least 48 h until electrophysiology studies.

## *Recording of Cytosolic [Ca²⁺]*

Cytosolic calcium transients were estimated in isolated cells using a confocal imaging system (Olympus Fluoview System version 4.2 FV300; TIEMPO) mounted on an upright Olympus microscope (IX71) as previously described (Au et al. 2009; Lee et al. 2010, 2011a; Ng et al. 2010). Briefly, cells were loaded with 1:1 (v/v) amount of 20% Pluronic-F127 (Invitrogen) and 5 μM Fluo-3 AM (Sigma-Aldrich) dissolved in DMSO with stock concentration of 5 mM for 45 min at 37 °C in Tyrode solution (Au et al. 2009). Calcium transients of single cardiomyocytes were recorded by Fluoview software with a temporal resolution of the line scan at 274 frames per second. Sarcoplasmic reticular (SR) calcium load was estimated from peak calcium release after the addition of caffeine (10 mM). All confocal calcium imaging experiments were performed within 48 h after isolation to minimize contamination of time-dependent changes in calcium handling properties in culture. Raw data of fluorescence intensity were recorded by area versus time mode (XYT) as a line plot: the calibration curve showed that there is a linear relationship between fluo-3 intensity recorded with calcium concentration up to 630 nM [$Ca^{2+}$] (Fig. 1.18).

Line-scan images were recorded in line versus time mode (XT). The data were then quantified as the background subtracted fluorescence intensity changes normalized to the background subtracted baseline fluorescence using Image J. Amplitudes, maximal upstroke, and decay velocity of calcium transients were analyzed by Clampfit version 9.2.0.09. (Axon Instruments, Foster City, CA, USA).

## RT-qPCR of Calcium Handling Proteins

**Table 1.3** Primers for reverse transcription-quantitative polymerase chain reaction (qPCR) of human cardiac-specific proteins and calcium handling proteins

| Gene | Direction | Sequence (5′ to 3′) |
|---|---|---|
| Nkx2.5 | Forward | TTCCCGCCGCCCCCGCCTTCTAT |
| | Reverse | CGCTCCGCGTTGTCCGCCTCTGT |
| α-MHC | Forward | GTTGGTGTTGGCTTGCTCCTC |
| | Reverse | ATCAAGGAGCTCACCTACCAG |
| β-MHC | Forward | TGGGGCTTTGCTGGCACCTCC |
| | Reverse | GCGGAGGAGCAAGCCAACACC |
| $Cav_{1.2}$ | Forward | TGACATCGAGGGAGAAAACT |
| | Reverse | ACATTAGACTTGACTGCGGC |
| NCX1 | Forward | TGTGCATCTCAGCAATGTCA |
| | Reverse | TGATGCCAATGCTCTCACTC |
| RYR2 | Forward | CGTTCTAACCAGCATCTCATC |
| | Reverse | CGAGCAATACAACCTGACC |
| SERCA2a | Forward | ACCCACATTCGAGTTGGAAG |
| | Reverse | CAGTGGGTTGTCATGAGTGG |
| CASQ2 | Forward | GAGCTTGTGGCCCAGGTCCT |
| | Reverse | GATCTCCACTGGGTCTTCAA |
| Junctin | Forward | GTAAAATGGCATCCCGAGAC |
| | Reverse | GGATGATGATGCCAGAGC |
| Triadin | Forward | TCAGTTGCTCCACACTGAGC |
| | Reverse | CCCATTTACAGACGGGAAAC |
| PLB | Forward | CTGCCAAGGCTACCTAAAAG |
| | Reverse | AGCTGAGCGAGTGAGGTATT |
| IP3R | Forward | GAAGAAACTACAGCACGTG |
| | Reverse | TTCTCCAGTAAAGCAGGTAA |
| GAPDH | Forward | AGCCACATCGCTCAGACACC |
| | Reverse | GTACTCAGCGCCAGCATCG |

*Nkx2.5* NK2 transcription factor related, locus 5, *α-MHC* alpha-myosin heavy chain, *β-MHC* beta-myosin heavy chain, *$Ca_v$ 1.2* L-type $Ca^{2+}$ channel, *NCX1* sodium calcium exchanger-1, *RyR2* ryanodine receptor-2, *SERCA-2a* sarcoplasmic reticulum calcium ATPase-2a, *CASQ2* calsequestrin-2, *PLB* phospholambam, *IP3R* inositol-1,4,5-triphosphate receptor, *GAPDH* glyceraldehyde 3-phosphate dehydrogenase

Germanguz's group showed that the EC coupling machinery of hESC-CMCs is different from that of the mature myocardium, mainly because of a dysfunctional SR $Ca^{2+}$ release capacity, and relies highly on extracellular $Ca^{2+}$ for contraction (Germanguz et al. 2011). Although underdeveloped SR function was found in hiPSC-CMC (Lee et al. 2011a), the EC coupling data remain minimal. Given the attractive application of hiPSC-CMCs in cell transplantation, this potential source for cell therapy should be further developed to attain functional compatibility with the adult myocardium. The contractility parameters have to be assessed in hiPSC-CMCs to further determine the maturity for interventional purposes.

## Excitation–Contraction (EC) Coupling Assessment

### *Measurement of Cardiac Cell Contractility by Video-Edge Detection*

The coupling of calcium homeostasis with mechanical action of EBs could be measured by a video-edge detector (Ren and Wold 2001; Germanguz et al. 2011). Briefly, spontaneously contracting areas with a diameter range of 0.5–1 mm were mechanically dissected out of the entire EB and adhered onto 30-mm-diameter glass slides. Subsequently, the contracting EBs were transferred to a chamber mounted on the stage of an inverted microscope and perfused with Tyrode's solution at a rate of 1–2 ml/min at 37 °C. EBs were field stimulated at different rates (0.25–0.5 Hz with 3-ms pulse duration) at suprathreshold voltages by means of platinum wires connected with an FHC stimulator (Brunswick, NE). The polarity of the stimulatory electrodes should be frequently reversed to avoid possible buildup of electrolyte by-products.

The EB was illuminated with red light, and a dichroic mirror (630-nm cutoff) in the emission path deflected the EB image to a video optical system (Crescent Electronics), which tracked the motion of the edges to determine the amplitude and velocity of cell shortening/relengthening. The motion signal was obtained at a rate of 60 Hz. Because contracting EBs have an irregular three-dimensional (3D) structure and, in contrast to adult ventricular myocytes do not shorten along a single axis of contraction, the output of the video-edge detector, which is expressed in arbitrary units, is not linearly indicative of the force of contraction. To characterize the contraction amplitude, the differences between minimal and maximal video cursor position ($L_{Amp}$) were measured in ten successive transients and averaged. Additionally, the maximal rate of contraction (dl/dt Contrac) and the maximal rate of relaxation (dl/dt Relax) were calculated (Germanguz et al. 2011).

# Limitations and Future Studies

Regarding calcium handling studies, fluorescence confocal recording of intracellular $Ca^{2+}$ release was not performed simultaneously with action potential recordings using a patch-clamp; thus, the chamber identity of individual cardiomyocytes (atrial, ventricular, and pacemaker myocytes) cannot be ascertained. Because atrial cells are known to lack t-tubules and they display an irregular internal transverse tubule system, the assessment of different cells may lead to nonrepresentative data of SR $Ca^{2+}$ handling function (Kirk et al. 2003; Shiels and White 2005). However, in the present study, the profiling of chamber-specific cells was similar in hiPSC-CMCs and hESC-CMCs; the two groups of cells are thus comparable to each other. Furthermore, whole-cell voltage clamp and simultaneous recording of intracellular $Ca^{2+}$ release can be used to characterize the retrograde and orthograde signaling between RyR2 and L-type calcium channels in cultured myotubes. The calcium current kinetics change in response to action potentials evoked by electrical stimulation will be investigated to determine the voltage dependence of CMC derived from the two stem cell types on calcium homeostasis (Eltit et al. 2011). In addition, to study EC coupling, video-edge detection of cell contraction activity is required to be investigated in junction with calcium fluorescence recording. The contractile activities mediated by the CICR mechanism to cardiac cell contraction would be investigated.

As an indicator of structural maturity of EC coupling, the t-tubule structure of hESC- or hiPSC-derived CMCs must be visualized by the membrane potential-sensitive dye, di-4-ANEPPS, or, by immunostaining of caveolin and L-type calcium channels, the proximity of the L-type calcium channel on the membrane to the RyR on SR can be revealed (Shiels and White 2005). In a t-tubule-rich cell, $Ca^{2+}$ influx from the cell periphery releases SR stores and then diffuses to the cell interior, which produces a homogeneous calcium spark (Louch et al. 2004; Lieu et al. 2009). The functional consequences of the structural advance will be accessed by line-scan imaging, which reveals temporal and spatial properties of cellular $Ca^{2+}$ flux in cardiomyocytes.

Calcium imaging was performed with Fluo-3 instead of ratiometric calcium indicators such as Indo-1 or Fura-2 because of the high sensitivity. However, using Fluo-3 precludes the possibility of measuring the absolute intracellular calcium concentration. Fura-2 is excited at the wavelengths 340 and 380 nm, and the ratio of the emissions at those wavelengths is directly correlated to the amount of intracellular calcium. The use of the ratio automatically cancels out confounding variables, such as variable dye concentration and cell thickness, making Fura-2 one of the most appreciated tools to quantify calcium levels.

# Conclusion

hiPSC-derived cardiomyocytes possess functional but immature SR. Although hiPSC-derived cardiomyocytes, avoiding the potential immune rejection and ethical issues peculiar to hESCs, hold promise for cardiac regeneration, the immature calcium handling properties of hiPSC-derived cardiomyocytes result in poor functional integration or lethal arrhythmia after cellular transplantation, thus limiting their potential therapeutic applications (Zhang et al. 2002; Liao et al. 2010). Therefore, it is highly desirable to develop strategies to drive functional maturation ex vivo. A standard protocol for phenotypic characterization of calcium handling in hiPSC-CMC may be important before usage in personalized in vitro models of cardiac tissue. In addition to clinical applications, the ability to generate patient-specific hiPSC-CMCs also provides an opportunity to develop novel in vitro models of cardiac disorders, which serve as a platform for investigation of arrhythmia mechanisms and for evaluation of the efficacy of pharmacological therapies. Taken collectively, the results of different studies in calcium handling of hiPSC-CMCs may have crucial implications for potential iPSC technology in basic and translational cardiac research.

# References

Ashrafian H, McKenna WJ, Watkins H (2011) Disease pathways and novel therapeutic targets in hypertrophic cardiomyopathy. Circ Res 109(1):86–96

Au KW, Liao SY, Lee YK, Lai WH, Ng KM, Chan YC, Yip MC, Ho CY, Wu EX, Li RA, Siu CW, Tse HF (2009) Effects of iron oxide nanoparticles on cardiac differentiation of embryonic stem cells. Biochem Biophys Res Commun 379(4):898–903

Bagutti C, Wobus AM, Fassler R, Watt FM (1996) Differentiation of embryonal stem cells into keratinocytes: comparison of wild-type and beta 1 integrin-deficient cells. Dev Biol 179(1):184–196

Bain G, Kitchens D, Yao M, Huettner JE, Gottlieb DI (1995) Embryonic stem cells express neuronal properties in vitro. Dev Biol 168(2):342–357

Balsam LB, Wagers AJ, Christensen JL, Kofidis T, Weissman IL, Robbins RC (2004) Haematopoietic stem cells adopt mature haematopoietic fates in ischaemic myocardium. Nature (Lond) 428(6983):668–673

Bers DM (2002) Cardiac excitation-contraction coupling. Nature 415(6868):198–205

Bers DM (2008) Calcium cycling and signaling in cardiac myocytes. Annu Rev Physiol 70:23–49

Bers DM, Despa S (2006) Cardiac myocytes $Ca^{2+}$ and $Na^+$ regulation in normal and failing hearts. J Pharmacol Sci 100(5):315–322

Bers DM, Perez-Reyes E (1999) Ca channels in cardiac myocytes: structure and function in Ca influx and intracellular Ca release. Cardiovasc Res 42(2):339–360

Blanco G, Mercer RW (1998) Isozymes of the Na-K-ATPase: heterogeneity in structure, diversity in function. Am J Physiol 275(5 Pt 2):F633–F650

Blaustein MP, Lederer WJ (1999) Sodium/calcium exchange: its physiological implications. Physiol Rev 79(3):763–854

Boheler KR, Czyz J, Tweedie D, Yang HT, Anisimov SV, Wobus AM (2002) Differentiation of pluripotent embryonic stem cells into cardiomyocytes. Circ Res 91(3):189–201

Brette F, Orchard C (2003) T-tubule function in mammalian cardiac myocytes. Circ Res 92(11):1182–1192

Campuzano V, Montermini L, Molto MD, Pianese L, Cossee M, Cavalcanti F, Monros E, Rodius F, Duclos F, Monticelli A, Zara F, Canizares J, Koutnikova H, Bidichandani SI, Gellera C, Brice A, Trouillas P, De Michele G, Filla A, De Frutos R, Palau F, Patel PI, Di Donato S, Mandel JL, Cocozza S, Koenig M, Pandolfo M (1996) Friedreich's ataxia: autosomal recessive disease caused by an intronic GAA triplet repeat expansion. Science 271(5254):1423–1427

Cerrone M, Napolitano C, Priori SG (2009) Catecholaminergic polymorphic ventricular tachycardia: a paradigm to understand mechanisms of arrhythmias associated to impaired Ca(2+) regulation. Heart Rhythm 6(11):1652–1659

Cheng H, Lederer WJ, Cannell MB (1993) Calcium sparks: elementary events underlying excitation-contraction coupling in heart muscle. Science 262(5134):740–744

Chopra N, Yang T, Asghari P, Moore ED, Huke S, Akin B, Cattolica RA, Perez CF, Hlaing T, Knollmann-Ritschel BE, Jones LR, Pessah IN, Allen PD, Franzini-Armstrong C, Knollmann BC (2009) Ablation of triadin causes loss of cardiac $Ca^{2+}$ release units, impaired excitation-contraction coupling, and cardiac arrhythmias. Proc Natl Acad Sci USA 106(18):7636–7641

Dani C, Smith AG, Dessolin S, Leroy P, Staccini L, Villageois P, Darimont C, Ailhaud G (1997) Differentiation of embryonic stem cells into adipocytes in vitro. J Cell Sci 110(Pt 11):1279–1285

Dolnikov K, Shilkrut M, Zeevi-Levin N, Danon A, Gerecht-Nir S, Itskovitz-Eldor J, Binah O (2005) Functional properties of human embryonic stem cell-derived cardiomyocytes. Ann N Y Acad Sci 1047:66–75

Dolnikov K, Shilkrut M, Zeevi-Levin N, Gerecht-Nir S, Amit M, Danon A, Itskovitz-Eldor J, Binah O (2006) Functional properties of human embryonic stem cell-derived cardiomyocytes: intracellular $Ca^{2+}$ handling and the role of sarcoplasmic reticulum in the contraction. Stem Cells 24(2):236–245

Drab M, Haller H, Bychkov R, Erdmann B, Lindschau C, Haase H, Morano I, Luft FC, Wobus AM (1997) From totipotent embryonic stem cells to spontaneously contracting smooth muscle cells: a retinoic acid and db-cAMP in vitro differentiation model. FASEB J 11(11):905–915

Eltit JM, Szpyt J, Li H, Allen PD, Perez CF (2011) Reduced gain of excitation-contraction coupling in triadin-null myotubes is mediated by the disruption of FKBP12/RyR1 interaction. Cell Calcium 49(2):128–135

Emanueli C, Lako M, Stojkovic M, Madeddu P (2005) In search of the best candidate for regeneration of ischemic tissues: are embryonic/fetal stem cells more advantageous than adult counterparts? Thromb Haemost 94(4):738–749

Fabiato A (1983) Calcium-induced release of calcium from the cardiac sarcoplasmic reticulum. Am J Physiol 245(1):C1–C14

Fabiato A (1992) Two kinds of calcium-induced release of calcium from the sarcoplasmic reticulum of skinned cardiac cells. Adv Exp Med Biol 311:245–262

Fedak PW (2008) Paracrine effects of cell transplantation: modifying ventricular remodeling in the failing heart. Semin Thorac Cardiovasc Surg 20(2):87–93

Fraichard A, Chassande O, Bilbaut G, Dehay C, Savatier P, Samarut J (1995) In vitro differentiation of embryonic stem cells into glial cells and functional neurons. J Cell Sci 108(Pt 10): 3181–3188

Fu JD, Li J, Tweedie D, Yu HM, Chen L, Wang R, Riordon DR, Brugh SA, Wang SQ, Boheler KR, Yang HT (2006) Crucial role of the sarcoplasmic reticulum in the developmental regulation of $Ca^{2+}$ transients and contraction in cardiomyocytes derived from embryonic stem cells. FASEB J 20(1):181–183

Gai H, Leung EL, Costantino PD, Aguila JR, Nguyen DM, Fink LM, Ward DC, Ma Y (2009) Generation and characterization of functional cardiomyocytes using induced pluripotent stem cells derived from human fibroblasts. Cell Biol Int 33(11):1184–1193

George CH, Higgs GV, Lai FA (2003) Ryanodine receptor mutations associated with stress-induced ventricular tachycardia mediate increased calcium release in stimulated cardiomyocytes. Circ Res 93(6):531–540

Gepstein L (2002) Derivation and potential applications of human embryonic stem cells. Circ Res 91(10):866–876

Germanguz I, Sedan O, Zeevi-Levin N, Shtrichman R, Barak E, Ziskind A, Eliyahu S, Meiry G, Amit M, Itskovitz-Eldor J, Binah O (2009) Molecular characterization and functional properties of cardiomyocytes derived from human inducible pluripotent stem cells. J Cell Mol Med 15(1):38–51

Germanguz I, Sedan O, Zeevi-Levin N, Shtrichman R, Barak E, Ziskind A, Eliyahu S, Meiry G, Amit M, Itskovitz-Eldor J, Binah O (2011) Molecular characterization and functional properties of cardiomyocytes derived from human inducible pluripotent stem cells. J Cell Mol Med 15(1):38–51

Gnecchi M, Zhang Z, Ni A, Dzau VJ (2008) Paracrine mechanisms in adult stem cell signaling and therapy. Circ Res 103(11):1204–1219

Guo T, Cornea RL, Huke S, Camors E, Yang Y, Picht E, Fruen BR, Bers DM (2010) Kinetics of FKBP12.6 binding to ryanodine receptors in permeabilized cardiac myocytes and effects on Ca sparks. Circ Res 106(11):1743–1752

Gyorke I, Hester N, Jones LR, Gyorke S (2004) The role of calsequestrin, triadin, and junctin in conferring cardiac ryanodine receptor responsiveness to luminal calcium. Biophys J 86(4):2121–2128

Gyorke S, Stevens SC, Terentyev D (2009) Cardiac calsequestrin: quest inside the SR. J Physiol 587(Pt 13):3091–3094

Hayashi M, Denjoy I, Extramiana F, Maltret A, Buisson NR, Lupoglazoff JM, Klug D, Takatsuki S, Villain E, Kamblock J, Messali A, Guicheney P, Lunardi J, Leenhardt A (2009) Incidence and risk factors of arrhythmic events in catecholaminergic polymorphic ventricular tachycardia. Circulation 119(18):2426–2434

He J, Conklin MW, Foell JD, Wolff MR, Haworth RA, Coronado R, Kamp TJ (2001) Reduction in density of transverse tubules and L-type Ca$^{2+}$ channels in canine tachycardia-induced heart failure. Cardiovasc Res 49(2):298–307

He JQ, Ma Y, Lee Y, Thomson JA, Kamp TJ (2003) Human embryonic stem cells develop into multiple types of cardiac myocytes: action potential characterization. Circ Res 93(1):32–39

Hilliard FA, Steele DS, Laver D, Yang Z, Le Marchand SJ, Chopra N, Piston DW, Huke S, Knollmann BC (2010) Flecainide inhibits arrhythmogenic Ca$^{2+}$ waves by open state block of ryanodine receptor Ca$^{2+}$ release channels and reduction of Ca$^{2+}$ spark mass. J Mol Cell Cardiol 48(2):293–301

Hole N, Smith AG (1994) Embryonic stem cells and hematopoiesis. In: Freshney RI, Pragnell IB, Freshney MG (eds) Culture of hematopoietic cells. Wiley Liss, New York, pp 235–249

Huser J, Lipsius SL, Blatter LA (1996) Calcium gradients during excitation-contraction coupling in cat atrial myocytes. J Physiol 494(Pt 3):641–651

Ikemoto N, Yamamoto T (2000) Postulated role of inter-domain interaction within the ryanodine receptor in Ca(2+) channel regulation. Trends Cardiovasc Med 10(7):310–316

Itzhaki I, Schiller J, Beyar R, Satin J, Gepstein L (2006) Calcium handling in embryonic stem cell-derived cardiac myocytes: of mice and men. Ann N Y Acad Sci 1080:207–215

Itzhaki I, Rapoport S, Huber I, Mizrahi I, Zwi-Dantsis L, Arbel G, Schiller J, Gepstein L (2011) Calcium handling in human induced pluripotent stem cell derived cardiomyocytes. PLoS One 6(4):e18037

Jaconi M, Bony C, Richards SM, Terzic A, Arnaudeau S, Vassort G, Puceat M (2000) Inositol 1,4,5-trisphosphate directs Ca(2+) flow between mitochondria and the endoplasmic/sarcoplasmic reticulum: a role in regulating cardiac autonomic Ca(2+) spiking. Mol Biol Cell 11(5):1845–1858

Jiang D, Xiao B, Yang D, Wang R, Choi P, Zhang L, Cheng H, Chen SR (2004) RyR2 mutations linked to ventricular tachycardia and sudden death reduce the threshold for store-overload-induced Ca$^{2+}$ release (SOICR). Proc Natl Acad Sci USA 101(35):13062–13067

Jiang D, Wang R, Xiao B, Kong H, Hunt DJ, Choi P, Zhang L, Chen SR (2005) Enhanced store overload-induced $Ca^{2+}$ release and channel sensitivity to luminal $Ca^{2+}$ activation are common defects of RyR2 mutations linked to ventricular tachycardia and sudden death. Circ Res 97(11):1173–1181

Jones PP, Jiang D, Bolstad J, Hunt DJ, Zhang L, Demaurex N, Chen SR (2008) Endoplasmic reticulum $Ca^{2+}$ measurements reveal that the cardiac ryanodine receptor mutations linked to cardiac arrhythmia and sudden death alter the threshold for store-overload-induced $Ca^{2+}$ release. Biochem J 412(1):171–178

Jung CB, Moretti A, Schnitzler MM, Iop L, Storch U, Bellin M, Dorn T, Ruppenthal S, Pfeiffer S, Goedel A, Dirschinger RJ, Seyfarth M, Lam JT, Sinnecker D, Gudermann T, Lipp P, Laugwitz KL (2011) Dantrolene rescues arrhythmogenic RYR2 defect in a patient-specific stem cell model of catecholaminergic polymorphic ventricular tachycardia. EMBO Mol Med 4: 180–191

Kapur N, Banach K (2007) Inositol-1,4,5-trisphosphate-mediated spontaneous activity in mouse embryonic stem cell-derived cardiomyocytes. J Physiol 581(Pt 3):1113–1127

Keeley EC, Boura JA, Grines CL (2003) Primary angioplasty versus intravenous thrombolytic therapy for acute myocardial infarction: a quantitative review of 23 randomised trials. Lancet 361(9351):13–20

Kehat I, Kenyagin-Karsenti D, Snir M, Segev H, Amit M, Gepstein A, Livne E, Binah O, Itskovitz-Eldor J, Gepstein L (2001) Human embryonic stem cells can differentiate into myocytes with structural and functional properties of cardiomyocytes. J Clin Invest 108(3):407–414

Keller GM (1995) In vitro differentiation of embryonic stem cells. Curr Opin Cell Biol 7(6):862–869

Kirk MM, Izu LT, Chen-Izu Y, McCulle SL, Wier WG, Balke CW, Shorofsky SR (2003) Role of the transverse-axial tubule system in generating calcium sparks and calcium transients in rat atrial myocytes. J Physiol 547(Pt 2):441–451

Knollmann BC (2009) New roles of calsequestrin and triadin in cardiac muscle. J Physiol 587(Pt 13):3081–3087

Kobayashi S, Bannister ML, Gangopadhyay JP, Hamada T, Parness J, Ikemoto N (2005) Dantrolene stabilizes domain interactions within the ryanodine receptor. J Biol Chem 280(8):6580–6587

Kobayashi S, Yano M, Suetomi T, Ono M, Tateishi H, Mochizuki M, Xu X, Uchinoumi H, Okuda S, Yamamoto T, Koseki N, Kyushiki H, Ikemoto N, Matsuzaki M (2009) Dantrolene, a therapeutic agent for malignant hyperthermia, markedly improves the function of failing cardiomyocytes by stabilizing interdomain interactions within the ryanodine receptor. J Am Coll Cardiol 53(21):1993–2005

Kobayashi S, Yano M, Uchinoumi H, Suetomi T, Susa T, Ono M, Xu X, Tateishi H, Oda T, Okuda S, Doi M, Yamamoto T, Matsuzaki M (2010) Dantrolene, a therapeutic agent for malignant hyperthermia, inhibits catecholaminergic polymorphic ventricular tachycardia in a RyR2(R2474S/+) knock-in mouse model. Circ J 74(12):2579–2584

Kong H, Jones PP, Koop A, Zhang L, Duff HJ, Chen SR (2008) Caffeine induces $Ca^{2+}$ release by reducing the threshold for luminal $Ca^{2+}$ activation of the ryanodine receptor. Biochem J 414(3):441–452

Kramer J, Hegert C, Guan K, Wobus AM, Muller PK, Rohwedel J (2000) Embryonic stem cell-derived chondrogenic differentiation in vitro: activation by BMP-2 and BMP-4. Mech Dev 92(2):193–205

Ladd AN (2007) Stem cell differentiation toward a cardio myocyte phenotype. In: Penn M (ed) Stem cells and myocardial regeneration. Humana Press, Totowas, pp 135–149

Lahat H, Pras E, Olender T, Avidan N, Ben-Asher E, Man O, Levy-Nissenbaum E, Khoury A, Lorber A, Goldman B, Lancet D, Eldar M (2001) A missense mutation in a highly conserved region of CASQ2 is associated with autosomal recessive catecholamine-induced polymorphic ventricular tachycardia in Bedouin families from Israel. Am J Hum Genet 69(6):1378–1384

Laitinen PJ, Brown KM, Piippo K, Swan H, Devaney JM, Brahmbhatt B, Donarum EA, Marino M, Tiso N, Viitasalo M, Toivonen L, Stephan DA, Kontula K (2001) Mutations of the cardiac ryanodine receptor (RyR2) gene in familial polymorphic ventricular tachycardia. Circulation 103(4):485–490

Lakatta EG (1992) Functional implications of spontaneous sarcoplasmic reticulum $Ca^{2+}$ release in the heart. Cardiovasc Res 26(3):193–214

Lee YK, Ng KM, Chan YC, Lai WH, Au KW, Ho CY, Wong LY, Lau CP, Tse HF, Siu CW (2010) Triiodothyronine promotes cardiac differentiation and maturation of embryonic stem cells via the classical genomic pathway. Mol Endocrinol 24(9):1728–1736

Lee YK, Ng KM, Lai WH, Chan YC, Lau YM, Lian Q, Tse HF, Siu CW (2011a) Calcium homeostasis in human induced pluripotent stem cell-derived cardiomyocytes. Stem Cell Rev 7:976–986

Lee YK, Ng KM, Lai WH, Man C, Lieu DK, Lau CP, Tse HF, Siu CW (2011b) Ouabain facilitates cardiac differentiation of mouse embryonic stem cells through ERK1/2 pathway. Acta Pharmacol Sin 32(1):52–61

Leenhardt A, Lucet V, Denjoy I, Grau F, Ngoc DD, Coumel P (1995) Catecholaminergic polymorphic ventricular tachycardia in children. A 7-year follow-up of 21 patients. Circulation 91(5):1512–1519

Lehnart SE, Mongillo M, Bellinger A, Lindegger N, Chen BX, Hsueh W, Reiken S, Wronska A, Drew LJ, Ward CW, Lederer WJ, Kass RS, Morley G, Marks AR (2008) Leaky $Ca^{2+}$ release channel/ryanodine receptor 2 causes seizures and sudden cardiac death in mice. J Clin Invest 118(6):2230–2245

Liao SY, Liu Y, Siu CW, Zhang Y, Lai WH, Au KW, Lee YK, Chan YC, Yip PM, Wu EX, Wu Y, Lau CP, Li RA, Tse HF (2010) Proarrhythmic risk of embryonic stem cell-derived cardiomyocyte transplantation in infarcted myocardium. Heart Rhythm 7(12):1852–1859

Lieu DK, Liu J, Siu CW, McNerney GP, Tse HF, Abu-Khalil A, Huser T, Li RA (2009) Absence of transverse tubules contributes to non-uniform $Ca(2+)$ wavefronts in mouse and human embryonic stem cell-derived cardiomyocytes. Stem Cells Dev 18(10):1493–1500

Lim PS, Chan EC, Lu TC, Yu YL, Kuo SY, Wang TH, Wei YH (2000) Lipophilic antioxidants and iron status in ESRD patients on hemodialysis. Nephron 86(4):428–435

Liu W, Yasui K, Opthof T, Ishiki R, Lee JK, Kamiya K, Yokota M, Kodama I (2002) Developmental changes of $Ca^{2+}$ handling in mouse ventricular cells from early embryo to adulthood. Life Sci 71(11):1279–1292

Liu N, Colombi B, Memmi M, Zissimopoulos S, Rizzi N, Negri S, Imbriani M, Napolitano C, Lai FA, Priori SG (2006) Arrhythmogenesis in catecholaminergic polymorphic ventricular tachycardia: insights from a RyR2 R4496C knock-in mouse model. Circ Res 99(3):292–298

Liu J, Fu JD, Siu CW, Li RA (2007) Functional sarcoplasmic reticulum for calcium handling of human embryonic stem cell-derived cardiomyocytes: insights for driven maturation. Stem Cells 25(12):3038–3044

Liu J, Lieu DK, Siu CW, Fu JD, Tse HF, Li RA (2009) Facilitated maturation of $Ca^{2+}$ handling properties of human embryonic stem cell-derived cardiomyocytes by calsequestrin expression. Am J Physiol Cell Physiol 297(1):C152–C159

Livrea MA, Tesoriere L, Pintaudi AM, Calabrese A, Maggio A, Freisleben HJ, D'Arpa D, D'Anna R, Bongiorno A (1996) Oxidative stress and antioxidant status in beta-thalassemia major: iron overload and depletion of lipid-soluble antioxidants. Blood 88(9):3608–3614

Lloyd-Jones D, Adams RJ, Brown TM, Carnethon M, Dai S, De Simone G, Ferguson TB, Ford E, Furie K, Gillespie C, Go A, Greenlund K, Haase N, Hailpern S, Ho PM, Howard V, Kissela B, Kittner S, Lackland D, Lisabeth L, Marelli A, McDermott MM, Meigs J, Mozaffarian D, Mussolino M, Nichol G, Roger VL, Rosamond W, Sacco R, Sorlie P, Thom T, Wasserthiel-Smoller S, Wong ND, Wylie-Rosett J (2010) Heart disease and stroke statistics–2010 update: a report from the American Heart Association. Circulation 121(7):e46–e215

Louch WE, Bito V, Heinzel FR, Macianskiene R, Vanhaecke J, Flameng W, Mubagwa K, Sipido KR (2004) Reduced synchrony of $Ca^{2+}$ release with loss of T-tubules: a comparison to $Ca^{2+}$ release in human failing cardiomyocytes. Cardiovasc Res 62(1):63–73

Maltsev VA, Rohwedel J, Hescheler J, Wobus AM (1993) Embryonic stem cells differentiate in vitro into cardiomyocytes representing sinusnodal, atrial and ventricular cell types. Mech Dev 44(1):41–50

Maltsev VA, Wobus AM, Rohwedel J, Bader M, Hescheler J (1994) Cardiomyocytes differentiated in vitro from embryonic stem cells developmentally express cardiac-specific genes and ionic currents. Circ Res 75(2):233–244

Marban E, Robinson SW, Wier WG (1986) Mechanisms of arrhythmogenic delayed and early afterdepolarizations in ferret ventricular muscle. J Clin Invest 78(5):1185–1192

Marks AR (2000) Cardiac intracellular calcium release channels: role in heart failure. Circ Res 87(1):8–11

Marty I, Faure J, Fourest-Lieuvin A, Vassilopoulos S, Oddoux S, Brocard J (2009) Triadin: what possible function 20 years later? J Physiol 587(Pt 13):3117–3121

Marx SO, Reiken S, Hisamatsu Y, Jayaraman T, Burkhoff D, Rosemblit N, Marks AR (2000) PKA phosphorylation dissociates FKBP12.6 from the calcium release channel (ryanodine receptor): defective regulation in failing hearts. Cell 101(4):365–376

Meissner G, Wang Y, Xu L, Eu JP (2009) Silencing genes of sarcoplasmic reticulum proteins clarifies their roles in excitation-contraction coupling. J Physiol 587(Pt 13):3089–3090

Menasche P, Hagege AA, Vilquin JT, Desnos M, Abergel E, Pouzet B, Bel A, Sarateanu S, Scorsin M, Schwartz K, Bruneval P, Benbunan M, Marolleau JP, Duboc D (2003) Autologous skeletal myoblast transplantation for severe postinfarction left ventricular dysfunction. J Am Coll Cardiol 41(7):1078–1083

Mery A, Aimond F, Menard C, Mikoshiba K, Michalak M, Puceat M (2005) Initiation of embryonic cardiac pacemaker activity by inositol 1,4,5-trisphosphate-dependent calcium signaling. Mol Biol Cell 16(5):2414–2423

Miller-Hance WC, LaCorbiere M, Fuller SJ, Evans SM, Lyons G, Schmidt C, Robbins J, Chien KR (1993) In vitro chamber specification during embryonic stem cell cardiogenesis. Expression of the ventricular myosin light chain-2 gene is independent of heart tube formation. J Biol Chem 268(33):25244–25252

Mineo C, Shaul PW (2003) HDL stimulation of endothelial nitric oxide synthase: a novel mechanism of HDL action. Trends Cardiovasc Med 13(6):226–231

Mineo C, Deguchi H, Griffin JH, Shaul PW (2006) Endothelial and antithrombotic actions of HDL. Circ Res 98(11):1352–1364

Moore JC, van Laake LW, Braam SR, Xue T, Tsang SY, Ward D, Passier R, Tertoolen LL, Li RA, Mummery CL (2005) Human embryonic stem cells: genetic manipulation on the way to cardiac cell therapies. Reprod Toxicol 20(3):377–391

Moorman AF, Vermeulen JL, Koban MU, Schwartz K, Lamers WH, Boheler KR (1995) Patterns of expression of sarcoplasmic reticulum Ca$^{2+}$-ATPase and phospholamban mRNAs during rat heart development. Circ Res 76(4):616–625

Moorman AF, Schumacher CA, de Boer PA, Hagoort J, Bezstarosti K, van den Hoff MJ, Wagenaar GT, Lamers JM, Wuytack F, Christoffels VM, Fiolet JW (2000) Presence of functional sarcoplasmic reticulum in the developing heart and its confinement to chamber myocardium. Dev Biol 223(2):279–290

Moschella MC, Marks AR (1993) Inositol 1,4,5-trisphosphate receptor expression in cardiac myocytes. J Cell Biol 120(5):1137–1146

Mummery C, Ward-van Oostwaard D, Doevendans P, Spijker R, van den Brink S, Hassink R, van der Heyden M, Opthof T, Pera M, de la Riviere AB, Passier R, Tertoolen L (2003) Differentiation of human embryonic stem cells to cardiomyocytes: role of coculture with visceral endoderm-like cells. Circulation 107(21):2733–2740

Mummery CL, Ward D, Passier R (2007) Differentiation of human embryonic stem cells to cardiomyocytes by coculture with endoderm in serum-free medium. Curr Protoc Stem Cell Biol Chapter 1: Unit 1F.2

Murry CE, Soonpaa MH, Reinecke H, Nakajima H, Nakajima HO, Rubart M, Pasumarthi KB, Virag JI, Bartelmez SH, Poppa V, Bradford G, Dowell JD, Williams DA, Field LJ (2004)

Haematopoietic stem cells do not transdifferentiate into cardiac myocytes in myocardial infarcts. Nature (Lond) 428(6983):664–668

Nadal-Ginard B, Kajstura J, Leri A, Anversa P (2003) Myocyte death, growth, and regeneration in cardiac hypertrophy and failure. Circ Res 92(2):139–150

Nakanishi T, Seguchi M, Takao A (1988) Development of the myocardial contractile system. Experientia (Basel) 44(11–12):936–944

Ng KM, Lee YK, Chan YC, Lai WH, Fung ML, Li RA, Siu CW, Tse HF (2010) Exogenous expression of HIF-1 alpha promotes cardiac differentiation of embryonic stem cells. J Mol Cell Cardiol 48(6):1129–1137

Ng KM, Lee YK, Lai WH, Chan YC, Fung ML, Tse HF, Siu CW (2011) Exogenous expression of human apoA-I enhances cardiac differentiation of pluripotent stem cells. PLoS One 6(5):e19787

Niggli E (1999) Localized intracellular calcium signaling in muscle: calcium sparks and calcium quarks. Annu Rev Physiol 61:311–335

Nygren JM, Jovinge S, Breitbach M, Sawen P, Roll W, Hescheler J, Taneera J, Fleischmann BK, Jacobsen SE (2004) Bone marrow-derived hematopoietic cells generate cardiomyocytes at a low frequency through cell fusion, but not transdifferentiation. Nat Med 10(5):494–501

Okabe S, Forsberg-Nilsson K, Spiro AC, Segal M, McKay RD (1996) Development of neuronal precursor cells and functional postmitotic neurons from embryonic stem cells in vitro. Mech Dev 59(1):89–102

Okita K, Ichisaka T, Yamanaka S (2007) Generation of germline-competent induced pluripotent stem cells. Nature (Lond) 448(7151):313–317

Orchard CH, Eisner DA, Allen DG (1983) Oscillations of intracellular $Ca^{2+}$ in mammalian cardiac muscle. Nature (Lond) 304(5928):735–738

Otsu K, Kuruma A, Yanagida E, Shoji S, Inoue T, Hirayama Y, Uematsu H, Hara Y, Kawano S (2005) $Na^+/K^+$ ATPase and its functional coupling with $Na^+/Ca^{2+}$ exchanger in mouse embryonic stem cells during differentiation into cardiomyocytes. Cell Calcium 37(2):137–151

Oudit GY, Trivieri MG, Khaper N, Husain T, Wilson GJ, Liu P, Sole MJ, Backx PH (2004) Taurine supplementation reduces oxidative stress and improves cardiovascular function in an iron-overload murine model. Circulation 109(15):1877–1885

Oudit GY, Trivieri MG, Khaper N, Liu PP, Backx PH (2006) Role of L-type $Ca^{2+}$ channels in iron transport and iron-overload cardiomyopathy. J Mol Med (Berl) 84(5):349–364

Park IH, Zhao R, West JA, Yabuuchi A, Huo H, Ince TA, Lerou PH, Lensch MW, Daley GQ (2008) Reprogramming of human somatic cells to pluripotency with defined factors. Nature (Lond) 451(7175):141–146

Passier R, Oostwaard DW, Snapper J, Kloots J, Hassink RJ, Kuijk E, Roelen B, de la Riviere AB, Mummery C (2005) Increased cardiomyocyte differentiation from human embryonic stem cells in serum-free cultures. Stem Cells 23(6):772–780

Paul-Pletzer K, Yamamoto T, Bhat MB, Ma J, Ikemoto N, Jimenez LS, Morimoto H, Williams PG, Parness J (2002) Identification of a dantrolene-binding sequence on the skeletal muscle ryanodine receptor. J Biol Chem 277(38):34918–34923

Paul-Pletzer K, Yamamoto T, Ikemoto N, Jimenez LS, Morimoto H, Williams PG, Ma J, Parness J (2005) Probing a putative dantrolene-binding site on the cardiac ryanodine receptor. Biochem J 387(Pt 3):905–909

Pegg W, Michalak M (1987) Differentiation of sarcoplasmic reticulum during cardiac myogenesis. Am J Physiol 252(1 Pt 2):H22–H31

Poindexter BJ, Smith JR, Buja LM, Bick RJ (2001) Calcium signaling mechanisms in dedifferentiated cardiac myocytes: comparison with neonatal and adult cardiomyocytes. Cell Calcium 30(6):373–382

Qin J, Valle G, Nani A, Nori A, Rizzi N, Priori SG, Volpe P, Fill M (2008) Luminal $Ca^{2+}$ regulation of single cardiac ryanodine receptors: insights provided by calsequestrin and its mutants. J Gen Physiol 131(4):325–334

Ren J, Wold LE (2001) Measurement of cardiac mechanical function in isolated ventricular myocytes from rats and mice by computerized video-based imaging. Biol Proced Online 3:43–53

Risau W, Sariola H, Zerwes HG, Sasse J, Ekblom P, Kemler R, Doetschman T (1988) Vasculogenesis and angiogenesis in embryonic-stem-cell-derived embryoid bodies. Development (Camb) 102(3):471–478

Roell W, Lewalter T, Sasse P, Tallini YN, Choi BR, Breitbach M, Doran R, Becher UM, Hwang SM, Bostani T, von Maltzahn J, Hofmann A, Reining S, Eiberger B, Gabris B, Pfeifer A, Welz A, Willecke K, Salama G, Schrickel JW, Kotlikoff MI, Fleischmann BK (2007) Engraftment of connexin 43-expressing cells prevents post-infarct arrhythmia. Nature (Lond) 450(7171): 819–824

Rohwedel J, Maltsev V, Bober E, Arnold HH, Hescheler J, Wobus AM (1994) Muscle cell differentiation of embryonic stem cells reflects myogenesis in vivo: developmentally regulated expression of myogenic determination genes and functional expression of ionic currents. Dev Biol 164(1):87–101

Rose O, Rohwedel J, Reinhardt S, Bachmann M, Cramer M, Rotter M, Wobus A, Starzinski-Powitz A (1994) Expression of M-cadherin protein in myogenic cells during prenatal mouse development and differentiation of embryonic stem cells in culture. Dev Dyn 201(3):245–259

Rosemblit N, Moschella MC, Ondriasa E, Gutstein DE, Ondrias K, Marks AR (1999) Intracellular calcium release channel expression during embryogenesis. Dev Biol 206(2):163–177

Rousseau E, Meissner G (1989) Single cardiac sarcoplasmic reticulum $Ca^{2+}$-release channel: activation by caffeine. Am J Physiol 256(2 Pt 2):H328–H333

Satin J, Itzhaki I, Rapoport S, Schroder EA, Izu L, Arbel G, Beyar R, Balke CW, Schiller J, Gepstein L (2008) Calcium handling in human embryonic stem cell-derived cardiomyocytes. Stem Cells 26(8):1961–1972

Sauer H, Theben T, Hescheler J, Lindner M, Brandt MC, Wartenberg M (2001) Characteristics of calcium sparks in cardiomyocytes derived from embryonic stem cells. Am J Physiol Heart Circ Physiol 281(1):H411–H421

Sedan O, Dolnikov K, Zeevi-Levin N, Leibovich N, Amit M, Itskovitz-Eldor J, Binah O (2008) 1,4,5-Inositol trisphosphate-operated intracellular Ca(2+) stores and angiotensin-II/endothelin-1 signaling pathway are functional in human embryonic stem cell-derived cardiomyocytes. Stem Cells 26(12):3130–3138

Seki S, Nagashima M, Yamada Y, Tsutsuura M, Kobayashi T, Namiki A, Tohse N (2003) Fetal and postnatal development of $Ca^{2+}$ transients and $Ca^{2+}$ sparks in rat cardiomyocytes. Cardiovasc Res 58(3):535–548

Shiels HA, White E (2005) Temporal and spatial properties of cellular $Ca^{2+}$ flux in trout ventricular myocytes. Am J Physiol Regul Integr Comp Physiol 288(6):R1756–R1766

Siu CW, Moore JC, Li RA (2007a) Human embryonic stem cell-derived cardiomyocytes for heart therapies. Cardiovasc Hematol Disord Drug Targets 7(2):145–152

Song LS, Sobie EA, McCulle S, Lederer WJ, Balke CW, Cheng H (2006) Orphaned ryanodine receptors in the failing heart. Proc Natl Acad Sci USA 103(11):4305–4310

Strubing C, Ahnert-Hilger G, Shan J, Wiedenmann B, Hescheler J, Wobus AM (1995) Differentiation of pluripotent embryonic stem cells into the neuronal lineage in vitro gives rise to mature inhibitory and excitatory neurons. Mech Dev 53(2):275–287

Takahashi K, Tanabe K, Ohnuki M, Narita M, Ichisaka T, Tomoda K, Yamanaka S (2007) Induction of pluripotent stem cells from adult human fibroblasts by defined factors. Cell 131(5): 861–872

Tateishi H, Yano M, Mochizuki M, Suetomi T, Ono M, Xu X, Uchinoumi H, Okuda S, Oda T, Kobayashi S, Yamamoto T, Ikeda Y, Ohkusa T, Ikemoto N, Matsuzaki M (2009) Defective domain-domain interactions within the ryanodine receptor as a critical cause of diastolic $Ca^{2+}$ leak in failing hearts. Cardiovasc Res 81(3):536–545

Terentyev D, Cala SE, Houle TD, Viatchenko-Karpinski S, Gyorke I, Terentyeva R, Williams SC, Gyorke S (2005) Triadin overexpression stimulates excitation-contraction coupling and increases predisposition to cellular arrhythmia in cardiac myocytes. Circ Res 96(6):651–658

The GUSTO Investigators (1993) An international randomized trial comparing four thrombolytic strategies for acute myocardial infarction. N Engl J Med 329(10):673–682

Thomson JA, Itskovitz-Eldor J, Shapiro SS, Waknitz MA, Swiergiel JJ, Marshall VS, Jones JM (1998) Embryonic stem cell lines derived from human blastocysts. Science 282(5391): 1145–1147

Tsien R, Pozzan T (1989) Measurement of cytosolic free $Ca^{2+}$ with quin2. Methods Enzymol 172:230–262

Uchinoumi H, Yano M, Suetomi T, Ono M, Xu X, Tateishi H, Oda T, Okuda S, Doi M, Kobayashi S, Yamamoto T, Ikeda Y, Ohkusa T, Ikemoto N, Matsuzaki M (2010) Catecholaminergic polymorphic ventricular tachycardia is caused by mutation-linked defective conformational regulation of the ryanodine receptor. Circ Res 106(8):1413–1424

Valdivia HH, Valdivia C, Ma J, Coronado R (1990) Direct binding of verapamil to the ryanodine receptor channel of sarcoplasmic reticulum. Biophys J 58(2):471–481

Velagaleti RS, Pencina MJ, Murabito JM, Wang TJ, Parikh NI, D'Agostino RB, Levy D, Kannel WB, Vasan RS (2008) Long-term trends in the incidence of heart failure after myocardial infarction. Circulation 118(20):2057–2062

Walter PB, Fung EB, Killilea DW, Jiang Q, Hudes M, Madden J, Porter J, Evans P, Vichinsky E, Harmatz P (2006) Oxidative stress and inflammation in iron-overloaded patients with beta-thalassaemia or sickle cell disease. Br J Haematol 135(2):254–263

Watanabe H, Knollmann BC (2011) Mechanism underlying catecholaminergic polymorphic ventricular tachycardia and approaches to therapy. J Electrocardiol 44(6):650–655

Wehrens XH, Lehnart SE, Huang F, Vest JA, Reiken SR, Mohler PJ, Sun J, Guatimosim S, Song LS, Rosemblit N, D'Armiento JM, Napolitano C, Memmi M, Priori SG, Lederer WJ, Marks AR (2003) FKBP12.6 deficiency and defective calcium release channel (ryanodine receptor) function linked to exercise-induced sudden cardiac death. Cell 113(7):829–840

Weitzer G, Milner DJ, Kim JU, Bradley A, Capetanaki Y (1995) Cytoskeletal control of myogenesis: a desmin null mutation blocks the myogenic pathway during embryonic stem cell differentiation. Dev Biol 172(2):422–439

Wiles MV, Keller G (1991) Multiple hematopoietic lineages develop from embryonic stem (ES) cells in culture. Development 111(2):259–267

Wobus AM, Wallukat G, Hescheler J (1991) Pluripotent mouse embryonic stem cells are able to differentiate into cardiomyocytes expressing chronotropic responses to adrenergic and cholinergic agents and Ca2+ channel blockers. Differentiation (Camb) 48(3):173–182

Wobus AM, Rohwedel J, Strubing C, Jin S, Adler K, Maltsev V, Hescheler J (1997) In vitro differentiation of embryonic stem cells. In: Klug E, Thiel R (eds) Methods in developmental toxicology. Blackwell, Berlin, pp 1–17

Xiao J, Tian X, Jones PP, Bolstad J, Kong H, Wang R, Zhang L, Duff HJ, Gillis AM, Fleischer S, Kotlikoff M, Copello JA, Chen SR (2007) Removal of FKBP12.6 does not alter the conductance and activation of the cardiac ryanodine receptor or the susceptibility to stress-induced ventricular arrhythmias. J Biol Chem 282(48):34828–34838

Yamanaka S (2007) Strategies and new developments in the generation of patient-specific pluripotent stem cells. Cell Stem Cell 1(1):39–49

Yang Z, Steele DS (2000) Effects of cytosolic ATP on spontaneous and triggered $Ca^{2+}$-induced $Ca^{2+}$ release in permeabilised rat ventricular myocytes. J Physiol 523(Pt 1):29–44

Yang HT, Tweedie D, Wang S, Guia A, Vinogradova T, Bogdanov K, Allen PD, Stern MD, Lakatta EG, Boheler KR (2002) The ryanodine receptor modulates the spontaneous beating rate of cardiomyocytes during development. Proc Natl Acad Sci USA 99(14):9225–9230

Yang L, Soonpaa MH, Adler ED, Roepke TK, Kattman SJ, Kennedy M, Henckaerts E, Bonham K, Abbott GW, Linden RM, Field LJ, Keller GM (2008) Human cardiovascular progenitor cells develop from a KDR+ embryonic-stem-cell-derived population. Nature (Lond) 453(7194): 524–528

Yokoo N, Baba S, Kaichi S, Niwa A, Mima T, Doi H, Yamanaka S, Nakahata T, Heike T (2009) The effects of cardioactive drugs on cardiomyocytes derived from human induced pluripotent stem cells. Biochem Biophys Res Commun 387(3):482–488

Young IS, Trouton TG, Torney JJ, McMaster D, Callender ME, Trimble ER (1994) Antioxidant status and lipid peroxidation in hereditary haemochromatosis. Free Radic Biol Med 16(3):393–397

Yousef ZR, Redwood SR, Marber MS (2000) Postinfarction left ventricular remodelling: where are the theories and trials leading us? Heart 83(1):76–80

Yu J, Vodyanik MA, Smuga-Otto K, Antosiewicz-Bourget J, Frane JL, Tian S, Nie J, Jonsdottir GA, Ruotti V, Stewart R, Slukvin II, Thomson JA (2007) Induced pluripotent stem cell lines derived from human somatic cells. Science 318(5858):1917–1920

Yuan Q, Fan GC, Dong M, Altschafl B, Diwan A, Ren X, Hahn HH, Zhao W, Waggoner JR, Jones LR, Jones WK, Bers DM, Dorn GW 2nd, Wang HS, Valdivia HH, Chu G, Kranias EG (2007) Sarcoplasmic reticulum calcium overloading in junctin deficiency enhances cardiac contractility but increases ventricular automaticity. Circulation 115(3):300–309

Zhang L, Kelley J, Schmeisser G, Kobayashi YM, Jones LR (1997) Complex formation between junctin, triadin, calsequestrin, and the ryanodine receptor. Proteins of the cardiac junctional sarcoplasmic reticulum membrane. J Biol Chem 272(37):23389–23397

Zhang YM, Hartzell C, Narlow M, Dudley SC Jr (2002) Stem cell-derived cardiomyocytes demonstrate arrhythmic potential. Circulation 106(10):1294–1299

Zhang J, Wilson GF, Soerens AG, Koonce CH, Yu J, Palecek SP, Thomson JA, Kamp TJ (2009) Functional cardiomyocytes derived from human induced pluripotent stem cells. Circ Res 104(4):e30–e41

Zwi L, Caspi O, Arbel G, Huber I, Gepstein A, Park IH, Gepstein L (2009) Cardiomyocyte differentiation of human induced pluripotent stem cells. Circulation 120(15):1513–1523

# About the Authors

The principal author, Dr. Lee Yee-Ki, is a Ph.D. graduate from the University of Hong Kong. Her research interests focus on hypertrophy signal–driven cardiac differentiation and stem cell–derived cardiomyocyte maturation. Her interest was further extrapolated to maturity of human induced pluripotent stem cell–derived cardiomyocytes (hiPSC-CMCs) by studying their calcium homeostasis compared with human embryonic stem cell (hESC) CMCs. She has authored altogether ten papers in the stem cell–related field, four as the primary author, in international journals such as *Molecular Endocrinology* and *Stem Cell Reviews*.

The corresponding author, Dr. Siu Chung-Wah, M.D., is currently an Associate Professor in the Division of Cardiology, the University of Hong Kong. His current research interests focus on atrial fibrillation, heart failure, and the development of a bio-artificial pacemaker for treating sick sinus syndrome, tissue engineering of myocardium, and human embryonic stem cells and induced pluripotent stem cells for regenerative medicine. He has published a total of 112 peer-reviewed articles of clinical research and basic science in international journals such as *JAMA*, *Circulation*, *JACC*, *Blood*, *Stem Cells*, *Stem Cell Reviews*, *Molecular Endocrinology*, *Journal of Metabolism*, and *Clinical Endocrinology*, and has written ten book chapters on cardiology.

L. Yee-Ki and S. Chung-Wah, *Calcium Handling in hiPSC-Derived Cardiomyocytes*,    49
SpringerBriefs in Stem Cells, DOI 10.1007/978-1-4614-4093-2,
© Lee Yee-Ki and Siu Chung-Wah 2012

# Index

L. Yee-Ki and S. Chung-Wah, *Calcium Handling in hiPSC-Derived Cardiomyocytes*,
SpringerBriefs in Stem Cells, DOI 10.1007/978-1-4614-4093-2,
© Lee Yee-Ki and Siu Chung-Wah 2012